In this delightful book, physicist 1 ... an intriguing journey towards faith. *Chasing Proof, Finding Faith* is funny, and also bracingly honest, as Rudelius shares his struggles with doubt and anxiety. I couldn't put it down.

ARD A. LOUIS, PhD, professor of theoretical physics at the University of Oxford

Chasing Proof, Finding Faith is an unflinching and courageous engagement with the most troubling intellectual objections to the Christian faith, combined with the author's thoughtful and compelling responses and profoundly vulnerable and endearing authenticity. All this is carried along by an incredibly interesting storyline. I've never read anything quite like it.

PAT MCLEOD, PhD, Cru chaplain at Harvard University, cofounder of the Mamelodi Initiative, and author of *Hit Hard*.

Imagine realizing that you're a Christian after you've taken a polygraph for a job interview at the NSA. Or picture yourself as a string theory scholar reasoning through the idea of miracles. Those are just two of the incredible stories in this journey to an unexpected faith in Jesus—a journey that includes ups and downs, doubt and faith, anxiety and peace; and a journey that invites you, the reader, to a more honest and hope-filled faith. I couldn't put it down and finished it (almost) in one sitting.

BETHANY JENKINS, vice president of Veritas Media

This raw, honest account captures the joys and the doubts that come with conversion and shows how silly it is to think that a commitment to cutting-edge science is incompatible with Christianity. Tom Rudelius is one who follows the evidence wherever it leads him—in theoretical physics, and in religious faith.

MOLLY WORTHEN, PhD, associate professor of history at the University of North Carolina at Chapel Hill, and *New York Times* contributor

String theorist Tom Rudelius maps his own journey into faith in a scientific manner. In *Chasing Proof, Finding Faith*, we don't discover an apologetic so much as a history of one man's fearlessness and discoveries—bit by bit, thought by thought, conclusion after conclusion, criticism after criticism, exploration after exploration, and answer after answer. Rudelius's memoir of his resistance, his struggles, his moments of insight, and his relationship with his twin brother provides a transparent story of conversion. One doesn't find simplicities and overcooked confidence; one finds a soul in search of God; and when found in Christ, a resting soul.

SCOT MCKNIGHT, PhD, professor of New Testament at Northern Seminary, and author of *A Church Called Tov*

Chasing Proof, Finding Faith

TYNDALE
REFRESH™

Think Well. Live Well. Be Well.

Chasing

± Proof

Finding
Faith

A YOUNG SCIENTIST'S
SEARCH FOR TRUTH
IN A WORLD OF
UNCERTAINTY

TOM RUDELIUS, PhD

Visit Tyndale online at tyndale.com.

Tyndale and Tyndale's quill logo are registered trademarks of Tyndale House Ministries. *Tyndale Refresh* and the Tyndale Refresh logo are trademarks of Tyndale House Ministries. Tyndale Refresh is a nonfiction imprint of Tyndale House Publishers, Carol Stream, Illinois.

For information about special discounts for bulk purchases, please contact Tyndale House Publishers at csresponse@tyndale.com, or call 1-855-277-9400.

Library of Congress Cataloging-in-Publication Data

A catalog record for this book is available from the Library of Congress.

ISBN 978-1-4964-7181-9

Printed in the United States of America

29	28	27	26	25	24	23
7	6	5	4	3	2	1

Contents

Foreword

ARON WALL

THE BOOK YOU ARE HOLDING is a remarkable one. There are lots of books out there promoting Christianity, by the type of people you might call *salesmen*. The goal of a salesman is to produce a watertight and squeaky-clean argument, to convince you that only one position is intellectually respectable and fully capable of servicing your needs. He is afraid to admit any weakness in his arguments. He is afraid that if he talks honestly about his own doubts and struggles, his audience will take it as a reason to reject the product he is promoting. If you want a book like that, I suggest you look elsewhere. My friend Tom Rudelius is not a salesman. But he is a person who cares deeply about what is real, in both scientific and religious contexts. And because of this, he is also unafraid to share his spiritual doubts and struggles, both before and after he became convinced that Christianity is objectively true.

This book is about a process of growth. Growth in learning that adult responsibility requires something more than just being "a good kid" who follows the rules well enough to be praised by authority figures. Growth in becoming a scientist. Growth in becoming more interested in faith. And a gradual process of becoming more thoughtful about how these various goals fit together.

Some people, when they talk about the supposed conflict between science and religion, imagine some mythical, archetypal battle in which Galileo and the Inquisition are forever facing off about whether or not the earth goes around the sun. Others have in mind some godless college professor seducing young college students away from the true faith with spurious arguments. Every now and then, life fits these simplistic stereotypes. But more often they blind us to what is actually going on.

One of the advantages of actually *being* a scientist (instead of idealizing "Science," with a capital *S*) is that one has a more realistic idea of what kinds of results the scientific method can actually show or prove. Similarly, actually *being* religious tends to give one a different perspective on what is most essential to faith in God—a rather different perspective than one might get just by arguing with atheists (or theists) online. Like Tom, I am both a practicing physicist and a practicing Christian. And I am convinced that there is no conflict between science and religion so long as we define both domains by what is *centrally* important to them. But a lot of people are more interested in arguing about peripheral matters: topics that shouldn't really be regarded as either established scientific theories *or* as core articles of faith.

This book is also the account of a sudden conversion, prompted by a dramatic crisis event. And like many Damascus Road events, the personal experiences that went into this event were unique. They don't follow the expected pattern. As C. S. Lewis said in *Surprised by Joy*, God is "very unscrupulous" in the methods he uses to recruit people for his Kingdom.

For example, if there's one thing modern skeptics and the Old Testament prophets agree on, it's that astrology is bunk. Generally speaking, we cannot predict political events on earth by looking up at the sky—and yet in the Christmas story we see God making an exception, using an astronomical event to bring the magi to the Christ child. These men were not Jews, but followers of the Zoroastrian religion. And yet God met these sages by starting where they were, and led them by a gradual journey to where he was. Now, I have no idea what sort of natural or supernatural event the Star of Bethlehem might have been. Perhaps it doesn't matter. What matters about the story, for our purposes, is this image: God using the magi's interest in the starry heavens to lead them on to the greater Light.

As a more recent example, when I was in graduate school I knew a friend who started identifying as a Christian after a six-hour religious-themed hallucination, brought on by smoking some marijuana. (Apparently, on rare occasions, this drug produces effects more reminiscent of LSD. Or so the internet tells me.) Now, to be clear, I would strongly recommend *against* anyone using drugs to gain religious insights, since most hallucinations probably have far more to do with one's own subconscious than anything divine. That's why I advised this friend to stick to more conventional spiritual practices in the future, such as Bible reading and attending church. I think that's a very sensible rule for human beings to follow, but it's not a rule by which God is bound. He can reach out and save people however he wants. As we see in the Gospels, when people came to Jesus, it didn't always happen in a way that is satisfying either to skeptics or religious traditionalists!

This book is not about drugs or astrology. But there is another pseudoscientific way to try to discover hidden truths, which plays a key role in Tom's story—namely, the polygraph test. Forget what you've seen in the movies: In reality the polygraph is a far cry from being an actual "lie detector" test. So I find it regrettable that US intelligence services still use this discredited and deceptive tool. But though I'm tempted to go on a long rant about this topic (and exceed my allotted word count), I'm not going to, because it would be a distraction from the most important part of Tom's story. I also won't steal Tom's thunder by explaining exactly how the polygraph fits into his life's narrative. You'll have to read the book to find that out.

The one aspect I wish to highlight is this: None of the government workers involved in the polygraph charade had any religious purpose in mind. (Nor would I want to be a part of any group that uses such recruiting techniques!) But that is not the ultimate truth of the situation. All along, it was God who was in the driver's seat. God was acting with love and mercy, first to reveal Tom to himself, and then to reveal himself to Tom.

As the Old Testament hero Joseph said to his brothers: "You meant evil against me, but God intended it for good."[1]

Many years before, they had sold Joseph into slavery, but the final outcome was that he became the ruler of Egypt, in order to save the lives of many people. I'm sure being betrayed and kidnapped didn't seem like a particularly good thing to the young Joseph at the time. But later, his perspective on this event was completely different, because he knew what had come out of it.

Similarly, a religious conversion changes our perspective on life. Before coming to Christ, our life may have seemed like a meaningless jumble of events. But afterwards, in retrospect, we can see how some of these events were orchestrated by God, who was reaching out to save us from our sins. There are still many unanswered questions. But now we are sharing life with another. Like getting married and starting a family, it changes the focus of the rest of our lives.

Perhaps you don't share this religious perspective. Perhaps you are coming to this book more as a skeptic. That's fine; this book is still written for you. Wherever you are in your spiritual journey, I hope you can appreciate the honesty and vulnerability displayed here, in this thought-provoking memoir.

ARON WALL, PhD, assistant professor of theoretical physics, University of Cambridge, and author of the blog *Undivided Looking*

Introduction

UNTESTABLE. *Irrational. Unscientific.*

I tend to hear the same objections to both my religious faith *and* my field of study, which is string theory. I chose string theory as my vocational pursuit because it is the best candidate for a "theory of everything" that would unify fundamental physics into a single coherent framework. String theory is controversial because it is difficult to verify empirically. Unlike most areas of science, where hypotheses are developed, tested by experiment, and refined by the scientific method, our knowledge of string theory comes almost exclusively from mathematical calculations.

I typically spend my days reading research papers written by other string theorists, writing research papers of my own, meeting with collaborators, performing calculations with paper and pen, and writing simple computer programs to perform calculations that are too difficult or time-consuming to do by hand. I almost never touch experimental data. I'm often not even aware of the latest experimental discoveries in physics until my dad (who is not a scientist) sends me an article from a popular science magazine such as *Scientific American* or *Wired*.

Truth be told, the absence of experiments in string theory is what sold me on it in the first place. I'm not opposed to experimental physics; I'm just not very good at it! At Cornell, during my undergrad years, my fellow physics majors amused themselves by watching me struggle unsuccessfully with the lab equipment. (They weren't quite as amused when I established the curve in our math classes.)

To some, string theory is antithetical to science, narrowly defined as the practice of the modern scientific method. This view is typically part of a larger narrative in which scientific progress has illuminated an otherwise dark, primeval, superstitious world. Accordingly, the scientific method serves as the grand arbiter of knowledge, and any conclusions drawn from scientific experiments are gospel truth and must never be questioned. Any other source of knowledge—if such a source even exists—is considered inferior.

But this view of science is misguided.[1] People living long before the scientific revolution understood that the natural world functions with uniformity and regularity. Back then, as today, experts in history, philosophy, mathematics, and law discerned truths about reality (though without many of the tools of modern science), shedding light in areas where the scientific method is relatively unhelpful. And though science has done some remarkable things, it has never allowed us to achieve absolute certainty.

That isn't a criticism of science. The truth is, we don't know anything with 100 percent certainty. We don't know with 100 percent certainty that there are other conscious minds besides our own. We don't know with 100 percent certainty that the world didn't begin five seconds ago with a built-in "past." The unvarnished truth is that our knowledge of reality isn't black-and-white. Our world is full of uncertainty. Science, as an enterprise, specializes in quantifying, minimizing, and navigating this uncertainty.

At its best, science does this incredibly well. Innovations such as statistical analysis, double-blind procedures, and repeated trials offer some areas of modern science an astonishingly high degree of certainty, justly meriting praise. In one of science's greatest triumphs, particle physicists have now measured the anomalous magnetic dipole moment of the electron to ten decimal places of precision, finding perfect agreement with theoretical predictions.

Other areas of science don't allow such a high degree of certainty in their experiments.

Not long ago, I was diagnosed with high cholesterol. Despite extensive online research of the subject, I'm still uncertain about whether I should eat egg yolks. When it comes to the field of nutrition, the complicated nature of the human body (plus the fact that subjects in nutrition experiments often lie about how much junk food they've eaten) makes it impossible to achieve the same level of certainty that particle physicists enjoy.

This uncertainty is not a failure of science. The true beauty of science is not that it correctly answers every question we ask of it, like some sort of divine oracle, but that it humbly admits when it doesn't know the answer, quantifying its uncertainty in terms of p-values and error bars and thereby encouraging future generations to seek answers with greater clarity. (Unfortunately, scientists don't always demonstrate the humility that science itself exhibits.)

Yet, while science pushes us to search for truth and clarity in the world around us, it doesn't excuse us from dealing with the inescapable uncertainties we encounter along the way. Nutritionists still must make dietary recommendations for the public to follow. I still must decide whether to toss the yolks from my breakfast omelet.

As a theoretical physicist, I make decisions about which calculations to attempt and which areas of research to pursue, even in the face of uncertainty. I settled on string theory not because I'm certain it is the correct theory of everything, but simply because it's the best option I've found, the clearest route toward an understanding of the fundamental laws of nature.

Likewise, as a human being, I make decisions about which spiritual path to follow.

I grew up in a largely nonreligious family and never thought much about God or theology until I was in college. When I began exploring the subject, guided by my twin brother, Steve, I soon found myself in yet another world of uncertainty. There was no end to the questions, doubts, and arguments for and against the existence of God. As in string theory, scientific experiments were not very helpful, but neither was my quest for truth a total shot in the dark. Some of the arguments for the existence of God carried weight; some of the arguments against the existence of God did too.

Yet, once I started exploring the subject of faith, there was no going back. Ignoring the fundamental realities underlying our sense of meaning, purpose, existence, and morality in order to persist in blissful ignorance would be a leap of faith of its own. Certainty wasn't an option, no matter where I turned. Instead, I had to figure out which option was the best one. To do that, I had to build a bridge of knowledge that would make the leap of faith as short as possible.

This book is the story of how I built that bridge, with help from many others. And it's the story of a strange new world called Christianity I found on the other side.

Part I

The Road to Faith

1

"TOM, I'VE DECIDED TO GET BAPTIZED."

When this message from my twin brother popped up on my phone one morning in May 2009, I was more than a little upset. I already felt betrayed by Steve's conversion to Christianity, and this was further salt in the wound.

"You've already been baptized," I messaged back, referring to the sprinkling we received at our grandparents' church in Minneapolis shortly after we were born. "These things don't wear off over time."

"We need to talk," Steve responded. "My baptism is on June 25. You should come, and I'll explain it then."

I agreed to go to Steve's baptism, but I wasn't enthusiastic about it. Ever since grade school, he and I had committed ourselves to the pursuit of two things and two things only: *academic excellence* and *sports*. Any departure from that path was a repudiation of the Rudelius Twin Way. I would feel similarly betrayed later that year, when he started dating (his future wife) for the first time.

Steve explained that his decision to get baptized wasn't out of religious duty, but rather because he had rejected God when we were growing up and now wanted to publicly announce his newfound faith as an adult. This bothered me too, not only because Steve was changing his chosen path, but because I was afraid he would now be looking down his nose at me.

"I think I'm a pretty moral person," I told him, "and I don't want you judging me."

Even more than the fear of Steve's judgment, I was afraid I was losing my brother. Growing up, he and I watched a lot of *The Simpsons*, so in my mind the typical Christian was someone like Ned Flanders, the Simpsons' quirky, nerdy, Bible-toting next-door neighbor.

"I'm happy for you becoming a Christian," I told him, "but you better not become Ned Flanders."

Religion was never part of my experience growing up. Because of my dad's job as a management consultant, we moved every few years—from Minnesota to Japan to England—eventually settling in Northern Virginia, where I remained through high school. But in all those years, I went to church only a handful of times, and only when I was with my grandparents.

Of course, that doesn't mean I didn't believe in God. After all, where else was I supposed to turn when I needed a miraculous intervention? I once had a minor leg injury and prayed that God would heal it in time for my indoor soccer game. However, when my leg actually got better, I didn't attribute it to supernatural intervention. I just figured I got lucky.

Back then, I would readily identify with Christianity if it was convenient for me. Once, in a high school anthropology class, the teacher asked us to raise our hands if we believed that everything in the Bible was literally true. My two friends in the class both raised their hands, so I did too. Now, I certainly had never read the entire Bible. I had read portions of Genesis in my tenth-grade English class as part of a unit on ancient literature from around the world, and that was about it. But I wanted to fit in, so I faked it.

My true feelings toward religion were much less positive. In a psychology class that same year, I read a story about a woman who had driven herself into a mental health crisis by trying to avoid behavior she had been taught was sinful. The assignment was to generate discussion questions for the class. My first question was "Does organized religion do more harm than good?"

I realize now that the question was rather misguided. To somehow lump all faith systems into a single category of "organized religion" would be like lumping Tylenol and heroin into a single category and asking whether drugs do more harm than good. To lump Jim Jones, L. Ron Hubbard, Mother Teresa, and Martin Luther King Jr. into the blanket category of "religion" would be silly. Yet, when I was in high school, my classmates and I did it all the time. I can't even remember my own answer to the question I posed. I guess it really didn't matter to me. At the time, all religion fit into an even broader category of things I didn't need to waste my time thinking about.

My functional agnosticism was broken in the rare moments when I truly needed outside help. When I was twelve, my mom's brother Jim passed away unexpectedly from a heart attack. And for a few brief moments, everyone in the family—parents, grandparents, aunts, and uncles—believed in heaven. It didn't matter that few of us professed belief in God or went to church on a regular basis. It was unthinkable that someone as young, kind, smart, and committed to his family as Jim could be anywhere but heaven, and so heaven had to be real. Regardless of our religious beliefs, we all prayed the week of Jim's passing.

For me, the sense of irreversible loss brought on by Jim's death was overwhelming. In a video game, when you get killed, you start the level over again as if nothing happened. But real life isn't a video game, and it hit me hard to realize that Jim would never bang on his drum set again or crack a home run for his company's softball team.

I cried at Jim's funeral when I started wondering whether he and I could play catch again someday when we were both in heaven. Yet, deep down, I had my doubts that such a place even existed.

2

AS A KID, I WANTED TO BE a professional baseball player when I grew up. Or a secret agent. It depended on whether I was out on the diamond or watching a James Bond movie. For the first few years of Little League, pitching in the major leagues didn't seem like such a far-fetched ambition. Dad taught Steve and me how to pitch at an earlier age than most of the other kids, so at the beginning of my young career, it seemed like I struck out just about every batter I faced.

Unfortunately, the other kids soon caught up, and by the time I was twelve, I was no longer a menace on the mound. In fact, I hadn't yet hit my adolescent growth spurt, so I was a lot smaller than the other kids. For love of the game, I continued playing through high school, but I never experienced much success beyond Little League.

I came much closer to realizing my dream of becoming a secret agent. If not for a failed polygraph test during college, I might have actually done it.

I'll come back to that story later.

I never expected to become a theoretical physicist. The teachers at my elementary school would probably be surprised as well: Starting in third grade, when they split the students into *gifted* and *general education* classes, Steve was placed in the gifted class, but I somehow missed the cut.

I always hated the idea that people might think I was stupid, and I hated losing to my twin brother even more. Steve and I did everything together, and competition was often at the center of it. Cards, computer games, sports, academics, it didn't matter. We even learned how to juggle by turning it into a competition. So when Mom interrupted my video game one summer day to tell me that Steve would be in the gifted class the next year, but I wouldn't, I exclaimed, "What!? But I'm smarter than he is!"

After she left the room, I buried my head in my pillow and cried, and I vowed I would someday prove the doubters wrong and show them I belonged with the gifted students.

"Someday" turned out to be sooner than I expected. When my fourth-grade teacher realized I was considerably ahead of the other kids in the general ed class, he and my mom convinced the higher-ups at the school to let me into the gifted program.

For the rest of my younger days, I excelled in just about every subject—well, except physics. Apart from eye exams, CPR certifications, and polygraphs, I think the only test I ever failed in my life was a fifth-grade test on the basic physics of light and geometric optics (the electromagnetic spectrum, reflection, refraction, etc.). I actually had the audacity to blame the teacher, writing at the top of my exam, "We never learned most of the material on this test." She responded with an equally snarky, "Well, you should have." Fortunately, I became a lot better at physics when I got older—though, to be honest, I still don't understand geometric optics as well as I should.

Dad played a key role in developing my math skills. He taught Steve and me the various arithmetic functions right before we learned about them in school, giving us an edge over the other kids. In fourth grade, for some reason, he taught me to factor quadratic polynomials of a single variable. You can imagine my excitement when—finally!—in seventh grade, my math teacher told us he was going to teach us how to factor quadratic polynomials. Then again, maybe you can't imagine it. I suppose most seventh-grade boys aren't dreaming about the day they'll learn factorization.

When I was a kid, most of my friends who went to church complained about how boring it was, so I never felt I was missing anything. It seemed that the point of religion was merely to teach people how to behave in generally moral ways, and I felt as if my parents did a pretty good job of that without church. Through their example, instruction, and guidance, Steve and I learned the virtues of honesty, integrity, and charity.

I remember Dad telling us a series of bedtime stories when we were young, about twin fox brothers named Franklin and Fredrick, who were clearly meant to represent Steve and me.

In each story, the foxes were placed in a difficult moral situation, in which they basically had to choose between what was right and what was easy. For instance, there was one story in which the twin foxes and their animal friends were playing in a dangerous place, and a friend was badly injured. What were the two brothers to do? If they left their friend alone, he might die. But if they went for help, everyone would know they had been trespassing and breaking the rules. In the end, they did the right thing and got help, and everyone was so relieved they had acted in time to save their friend that they didn't even punish them for trespassing. And so we learned that friends are more important than reputations, and that—at least sometimes—honesty is rewarded.

Mom often let us learn through experience. One time, when Steve and I were about ten, we were casually throwing rocks at the brick concession stand at the Little League field. A woman saw us and thought we were trying to vandalize the drinking fountain on the side of the building. Somehow she found our mom and tattled on us. The two of them approached us just as we were about to throw our next rocks.

"This woman says you two were trying to break the water fountain," said Mom. "Were you?"

"No," I said, "that wasn't us."

Technically, it wasn't a lie. We weren't *trying* to hit the water fountain; we were aiming for certain bricks. Certainly I could have hit the water fountain with a rock from ten yards away if I'd wanted to. After all, I was going to be a professional baseball player. But why did that woman have to tattle on us to Mom? Why didn't she just tell us to stop?

"That's what I thought," Mom said to the woman. "*My* children would never do something destructive like that. *My* children are good kids."

It was a classic parenting move on Mom's part. The woman had clearly caught us red-handed doing something we weren't supposed to, but Mom defended us and allowed our guilty consciences to be our punishment.

We always knew our parents loved us, but they expressed it in different ways. To the best of my recollection, I was twenty-five when my dad first gave me a hug. Before that, a firm handshake was his way of saying good-bye. Growing up, I got the good-bye handshake once a week, as Dad would typically fly out on Monday, spend the week in another state or country on business, and fly home for the weekend on Thursday or Friday. He was doing what he had to do to support the family, but I certainly didn't see it that way at the time. Instead, I responded to his lack of warmth and frequent absences by freezing him out. When he would arrive home late Thursday night and yet wake up early Friday morning to wish us good luck at school, I would often respond with silence or a terse "thanks."

That said, when Dad *was* home, he was always willing to drop whatever he was doing to play with us: cards, football, tennis, golf, baseball, or anything else Steve and I wanted to do. Despite a demanding job, he somehow managed to spend time with his sons, and he never once said he wasn't in the mood to play with us.

When it came to physical affection, Mom was only slightly more expressive than Dad. The first time I can remember a hug from her was my senior year of high school, as she was leaving to fly home for her father's funeral. That week, as our house inevitably fell into disarray in her absence, I realized how much I took her for granted. Mom has an MBA from Northwestern, but when Steve and I were born, she sacrificed her career to stay home with us instead. Every morning, even through high school, she woke up at 6:00 a.m. to cook breakfast and pack our lunches. She was even our alarm clock—gently ruffling our sheets when it was time to wake up. She never once took a sick day or accidentally slept in. If Mom didn't wake us up at the usual time, it had to be a snow day, not an error on her part.

Steve and I mirrored our parents' ways of showing affection in our own relationship. We never hugged or talked about our love for each other, but there was an unbreakable bond between us—whether we acknowledged it or not.

After seeing our unique ability to understand each other, an observer might have concluded that Steve and I had some sort of paranormal "twin telepathy." But we didn't need telepathy: After years of shared experiences, we naturally developed a common vocabulary of inside jokes and references that allowed us to communicate in ways that no one else in the room would understand.

Once, while playing a word association game against some of our friends, Steve offered the word *hovercraft*, to which I correctly responded, "Lagoon." To the average person, there is no conceivable link between a hovercraft and a lagoon, so our friends assumed we were cheating. But for twin brothers who grew up playing the Pirate Lagoon level in Nintendo's Diddy Kong Racing—a level that can only be performed in a hovercraft, as opposed to a car or a plane—the connection was obvious.

3

THE BASIC MORALITY I LEARNED at home, combined with the loving environment in which I was raised and educated, turned me into a generally good guy, at least to all appearances.

I attended an elite magnet school, Thomas Jefferson High School for Science and Technology, in Fairfax County, Virginia, where I got good grades and played varsity baseball. I volunteered in the community—at least when it was required by the National Honor Society. I didn't smoke, drink, or use other drugs. I drove approximately the speed limit and even used my turn signals. My high school friends were all good kids as well. We bonded over our mutual love for sports and played football or basketball just about every day during lunchtime or after school, and we stayed out of trouble.

But if you kept digging, you'd find some dirt on me. I illegally downloaded music because I didn't feel like paying for it. I snuck into high school football games without a ticket for the same reason. More than anything, I did whatever it took to build and maintain my reputation. I would lie or meander my way around the truth to preserve the image I had built of myself: Tom Rudelius, the brilliant, athletic, clean-cut, all-American kid. Any evidence to the contrary was filtered out of my consciousness.

Morality was important to me, but it was a means to an end, rather than something to pursue on its own merits. The ultimate goal was *respect*—from my friends, neighbors, peers, teachers, and coaches. As long as nobody could see my flaws, I was unconcerned. At the same time, if my talents were invisible to everyone else, what was the point of having them? As a result, I was extremely cocky in the classroom and made sure the other kids knew I was smarter than they were. I once took a math test in crayon, just to show how easy it was for me.

Every once in a while, I caught a glimpse of the shallowness of my worldview. One day, some of the students in my homeroom class were having a discussion about the Make-A-Wish Foundation, which admirably seeks to fulfill the dreams of boys and girls with life-threatening medical conditions. One student interjected a contrary viewpoint into the discussion.

"What's the point?" he asked. "They're just going to die anyway."

I joined in the chorus of boos.

"How could you say such a thing!?" someone asked him. "That's terrible!"

Looking back, I now realize that, for all our self-righteous denunciation of the student's objection, at no point did anyone try to address his question seriously. Nobody had a good answer. We had been taught to operate from a naturalistic perspective, in which morality was merely a social construct, determined by the approval or disapproval of society. Arguments appealing to any higher power were forbidden.

With these ground rules, we had nothing to offer beyond a subjective, emotional response to our fellow student's statement—essentially that we didn't approve of what he was saying. Objectively, his claim was true: Sooner or later, these children would die—like all of us—their dreams swallowed up in a vast expanse of nothingness whether or not they had been fulfilled.

But my high school classmates and I never made the connection between our professed ideologies and our actions. The student who had insulted the Make-A-Wish Foundation wasn't living a life that screamed, "What's the point if I'm just going to die anyway?" Instead, like just about everyone else in the school, he was a decent guy and a solid student, and he went on to a top university.

Over time, the inconsistencies began to bother me. I grew curious about religion and the meaning of life, even asking my mom once or twice if we could go to church.

"Maybe," she always said, but we never did. And any desire to know more about God was quickly replaced once I moved on to college.

4

AFTER A STELLAR HIGH SCHOOL academic career, I was stunned to be rejected by Princeton, my top college choice. Instead, I packed my bags for Cornell as Steve headed off to Illinois to attend Northwestern. In high school, I had easily breezed to the top of my classes, and now I was determined to outwork and outperform my college classmates and prove I was light-years ahead of them. If I was assigned a problem set due on Friday, I would work at it for hours on end to have it done by Tuesday instead.

This relentless study strategy paid off in the classroom. Starting off my freshman year, I took an honors physics course that was intended for the top sophomore physics majors. At first, I struggled to understand the material, and

I thought about dropping the course—along with physics as a major—before I even took my first midterm (or prelim, as they call them at Cornell).

Then the tests came back. The mean was something like a 70, the high score was a 98, and I had a 97. It's a good thing I didn't drop that course.

Beyond academics, though, my life was rather depressing that first semester. Mom and Dad separated when Steve and I went off to college, and they finalized their divorce two years later. I didn't have many friends, and I was known in the dorm as "the most mysterious character on the floor." While most other students were in the common area socializing or out drinking with friends at a frat party, I was in my room, alone, studying. If I needed a break, I went to the gym and lifted weights (alone) or watched some TV or YouTube clips (alone).

I went through a phase of watching comedy acts by guys like Chris Rock, Lewis Black, and especially George Carlin. He was known for his unapologetic mockery of just about everything, but he particularly liked to take jabs at religion—especially Christianity. Here's one of his more famous quotes from a video I watched around that time:

> Think about it: Religion has actually convinced
> people that there's an invisible man living in the sky,
> who watches everything you do every minute of
> every day. And the invisible man has a special list of
> ten things he does not want you to do. And if you do
> *any* of these ten things, he has a special place full of

fire and smoke and burning and torture and anguish, where he will send *you* to live and suffer and burn and choke and scream and cry, forever and ever, til the end of time.

   But he *loves* you. He loves you. He loves you and he *needs money!*[1]

Even all these years later, I still laugh when I hear that. Now that I know more about what Christianity actually teaches, I would contend that the only accurate depiction of Christian theology in Carlin's bit is, "But he loves you." But I didn't know that back then. And when your life is a mixture of challenging studies, family disintegration, and overwhelming isolation, it feels good to be able to laugh at someone.

Steve and I texted each other regularly about pro sports our first year of college, but as usual, we avoided more serious topics. I didn't tell him about the problems I was having at Cornell, and remarkably we never once discussed our parents' separation. I assumed he was buried in his schoolwork at Northwestern, experiencing similar feelings of grief and isolation, and developing his atheistic worldview in the university environment. So I was completely floored when he told me he had decided to become a Christian. And I was floored again when he invited me to his baptism.

5

STEVE WAS BAPTIZED IN LAKE MICHIGAN, just off the Northwestern University campus. Mom and I drove down from her home in Minnesota, where I was spending my first summer home from Cornell.

The entire weekend was a montage of new experiences. I went to Steve's Sunday morning church service, which was held in a movie theater and featured electric guitars and a drum set. While the band played, a young woman in the row in front of me raised her hands in the air and seemed transfixed by the music.

This is not my grandparents' church service, I said to myself.

After the service, I met Steve's Christian friends, who might have been the first serious Christians my own age I had ever come across. Worried they would judge me for my lack of religiosity, I tried to disguise how out of place I felt. But I don't think they were buying my act. Steve must have already told them about his family, so they probably knew that I never went to church.

Finally, it was time for the baptism. As Mom, Steve, and I drove to a beach near campus, I avoided the topic of religion by asking Steve if he knew how cold the water would be this time of year.

"Pretty cold," he said.

When we arrived, some of Steve's friends had already gathered. The pastor invited Steve to share a bit of his "testimony," the story of how he had decided to become a Christian.

Steve began his talk in classic fashion: "On *The Simpsons*," he said, "God is the only character with five fingers—the rest all have four. Growing up, this was just about all I knew about God—that he has five fingers."

He went on to explain how a friend named Matt had shared with him about Christianity, and how he had gradually come to appreciate the sacrifice Jesus had made for him by dying on the cross. He said that school and sports had taken the ultimate place in his life, a place he now saw should be reserved only for God. He described how tearing his ACL playing pickup basketball had made him reevaluate his priorities. He shared how his overwhelming, obsessive fear of death had been alleviated by the knowledge that "whether we live or die, we belong to the Lord."[2]

When Steve was done, the pastor made a few remarks about baptizing Steve "in the name of the Father, the Son, and the Holy Spirit," and then dunked him under the chilly waters of the lake.

After the baptism, I hung out with Steve and his friends. I was pleasantly surprised to find they weren't haughty or judgmental, as I had feared. They had a long-running inside joke about the Book of Zion, an imaginary book of the Bible that they would cite in support of whatever they wanted.

"Sorry, Steve, your baptism doesn't count unless you chug a root beer beforehand," one of them said. "It's right there in the Book of Zion, chapter 15, verse 4."

"Aw, c'mon, God," Steve responded in mock frustration. "Let me in the club!"

I soon realized they were a bunch of Christian nerds—cracking jokes about nonalcoholic beverages and make-believe books of the Bible. This didn't help to alleviate my fears that Steve was about to become a real-life version of Ned Flanders, but at the same time I appreciated his friends' lighthearted attitude toward their faith.

That evening, Steve and I finally had some time to ourselves, and we eventually got around to the topic of the baptism. I asked him why he had started looking into religion in the first place. I had been drifting further and further toward atheism, thanks in part to the comedic influence of George Carlin and others, and I thought at the time that our shared disdain of religion would give Steve and me something else to bond over, aside from Minnesota Twins' baseball, *Seinfeld*, and the band Matchbox 20. I was baffled when he told me his journey to faith had begun with five simple words: "I don't believe in evolution."

"*Whoa, really?*" I asked, incredulous. I had been reluctant to broach the subject of religion at all, but Steve immediately piqued my interest with that introduction.

"Tell me more."

During his first semester at Northwestern, Steve said, he was hanging out with some guys from his dorm floor. Every summer, incoming freshmen are given a book to read before they arrive on campus, and that year the assigned book was about Charles Darwin. At one point, they started talking about the reading and the theory of evolution, and from there the conversation turned to ridiculing the anti-evolution movement.

Then Steve's friend Matt spoke up and said, "I don't believe in evolution," and everyone laughed.

Everyone, that is, except Matt.

"Wait, are you serious?" someone asked him.

"Yeah," Matt said. "The Bible teaches that evolution isn't true, and I believe the Bible."

"Stop. Hold on. *What?*" Steve had heard that people like this existed, but they weren't supposed to exist here, at a great school like Northwestern. And Matt was no intellectual slouch—he was a straight-A student studying engineering.

They all still thought Matt was pulling their leg, but no, he was serious. They were face-to-face with an intelligent, kind, polite, humble, absolutely ridiculous evangelical Christian.

The guys started grilling Matt about creation, evolution, and Christianity in general. As Matt explained his worldview and laid out his argument, Steve was surprised: He didn't find all of Matt's points compelling, and he certainly wasn't about to abandon his belief in evolution, but he was nonetheless impressed by how much Matt knew about these topics—which was far more than Steve did at the time.

The conversation continued for a while, and one by one the guys left to get back to their schoolwork. Eventually, it was just Steve and Matt.

"I want to talk more about this with you," Steve said.

"Yeah, for sure!" said Matt. "Tell you what: If you want to know more, why don't I give you a couple of books you can take a look at before we talk?"

Steve agreed, and Matt gave him a copy of the New Testament and a book called *Letters from a Skeptic* by Gregory and Edward Boyd. Little did he know the impact those books would have on Steve's life—and, not too long afterward, on mine as well.

I was surprised that Steve had taken an interest in what Matt said about evolution. Heading into his first semester of college, Steve was probably more hostile toward religion than I was. When we were applying to schools, he had ridiculed even the suggestion of going to a school with a religious affiliation.

As Steve shared his story with me, I was still trying to understand what had led him to want to know more about religion in the first place. In his testimony at the baptism, Steve had mentioned how he had spent much of the previous summer obsessing about death—specifically his own death, about ceasing to exist. I was surprised to hear this because he had never said anything to me about it. In fact, I had never even noticed that something was bothering him.

Now I felt like I had somehow failed in my duty as a brother to Steve by not helping him through this time of anxiety. I asked him when he had started having these fears, and he told me that they'd occasionally bothered him even as a child.

He told me about a time, when we were about twelve, that the thought of his own demise had kept him awake into the wee hours of the morning. Finally, unable to sleep, he had knocked on our parents' bedroom door. When Mom asked him what was wrong, he said, "I can't stop thinking about death."

"Aww, honey," Mom had replied soothingly. "You don't need to worry about that for a long time."

In a sense, her answer helped, Steve said, but it didn't resolve his problem. Whether he lived another five minutes or eighty years, death would eventually come for him. And beyond that . . . what exactly? Pure nothingness?

One day, after our senior year of high school, when Steve's obsession was at its peak, he tried looking up different views of the afterlife in search of some sort of reassurance. The various religions he explored seemed nice in some respects, but he wasn't convinced about nirvana, heaven, or reincarnation. It just seemed like religious mumbo jumbo. When he finally stumbled upon naturalism, he felt as if he had found a winner. Naturalism rejects any sort of reality beyond the physical world, looking to science as the best way to understand existence and how things work. And to the extent that, according to physics, energy is neither created nor destroyed, the energy in our bodies will continue to exist in the universe after we die, but not in some super-natural way.

Within a day or so, Steve realized that naturalism's approach to death, despite an impressive effort from its public relations team, was ultimately unsatisfying as well. When all is said and done, death means an end of consciousness, an end of the person known as Steve Rudelius.

So perhaps it wasn't so surprising that Steve found Matt's words so intriguing. On the one hand, he thought religion was a bunch of unscientific gobbledygook. On the other hand, he found atheism to be existentially unsatisfying. Now here was Matt, an engineering student and a Christian, claiming to offer a third way: religion for the intelligentsia. Steve wasn't convinced, but he wanted to hear more.

A little while later, Matt introduced Steve to a guy named Drew, a recent Northwestern graduate who now worked in a Christian campus ministry. Drew was the kind of guy who got along with everyone, who could walk into a room of strangers and instantly make people feel as if they'd known him for years.

Drew encouraged Steve to share a bit about himself, and then asked him, "How familiar are you with the Bible?"

"Not familiar at all," Steve said. Truthfully, most of what he knew about the Bible came from watching *The Simpsons*.

"Would you mind if I read a bit of it with you?" Drew asked him.

When Steve agreed, Drew opened up his Bible to the second chapter of the Gospel of John, where Jesus is attending a wedding in a town called Cana and the hosts run out of wine. This would have meant the end of the party, and a great social embarrassment for the hosts. Jesus saves the party by turning water into wine—and high-quality wine, at that.

This was quite surprising to Steve. He had never really gotten into alcohol, but it was still surprising to see it being celebrated in the Bible. In his mind, Christians were the sort of people who would turn wine into water if they could, not the other way around. They were the sort of people who made up rules to keep others from having fun, not the sort of people who had fun and threw parties and celebrated life. Yet here was Jesus, using apparently miraculous powers to keep a party going.

Drew asked Steve some questions about the passage and brought his attention to John 2:6, which describes the large stone water jars used for ceremonial washing—each one holding about twenty or thirty gallons—that Jesus used to make the wine.

"Why do you think John goes to all this trouble to tell us about the jars?" Drew asked.

Steve noted the size of the jars—it was clearly a *lot* of wine.

"Good point," said Drew. "But notice also what they're used for—these aren't made for holding wine, they're made for ceremonial washing. Imagine the sort of scandal it would have caused if the religious leaders found out about it—their sacred jars being used for wine! To Jesus, religious traditions were less important than people's hearts."

This was an important revelation to Steve—and an intriguing idea to me as well. When we were growing up, Christianity to us was primarily a list of dos and don'ts, and a set of religious rituals to be followed whether or not they made sense. Yet here was Jesus, brazenly flouting the religious traditions of his day.

To be fair, as I've grown in my Christian faith, I've begun to appreciate the heart behind many of the traditions, which I once viewed as mindless conformity. But when Steve and I first started exploring Christianity, it was a big deal to realize that we didn't have to learn all the jargon and rituals and rules. We could go straight to having "a relationship with God," as Drew called it.

Drew then showed Steve some other places in the Bible where wine, feasts, and weddings are mentioned, explaining how these seemingly distinct stories connected with one another to produce a larger narrative arc. At the end of their meeting, Steve still had a lot of questions and doubts—it was pretty hard to imagine that someone could actually turn water into wine, for one thing. But there was a lot more in the Bible than he ever would have noticed on his own—it was really quite entertaining, surprising, and rich—and it was starting to become clear to him that Christianity wasn't quite the religion he had imagined.

6

OVER THE NEXT SEVERAL MONTHS, Steve met with Drew several more times, and his appreciation for the Bible and Christianity continued to grow. One day, Drew and Matt invited him to join them for a Bible study with some of the other students on the floor.

In retrospect, Steve said, it was just about the worst introduction to Christian community you could imagine. Two students took over the conversation and bickered endlessly about obscure, irrelevant topics. When Steve introduced himself and talked about how he was new to Christianity and didn't know very much about it, one of the students replied, "Well then, I'm not sure you should even be here." Drew and Matt quickly assured Steve that he indeed belonged in the group.

Though Steve's first small-group experience was far from ideal, he did enjoy his conversations with Matt and Drew. He also had more free time available to explore Christianity, because he had torn his ACL playing basketball and would be unable to run, jump, or participate in sports for nearly a year. For someone who had competed in track all four years in high school, that was a devastating blow.

Christianity offered Steve an opportunity to root his identity in something other than sports and academics, and to gain a sense of purpose. Instead of lifting weights and working to improve his time in the 100 meters, he turned his attention to growing in character and exploring a relationship with God.

The first time he and Drew had met, Drew had shown him John 1:12-13: "To all who believed him and accepted him, he gave the right to become children of God. They are reborn—not with a physical birth resulting from human passion or plan, but a birth that comes from God."

"Just as Jesus turned the water into wine," Drew had said, "he can turn us into children of God."

Steve came to believe that Jesus could do more than just improve his *life*—he could actually improve his character as well. Like me, Steve had always been a good kid—he got good grades, he didn't smoke or drink, he was nice and well-liked. But pretty quickly he realized that Christianity is about more than just clean living. In the Bible, he discovered that Jesus spent a lot less time admonishing wayward "sinners" and a lot more time criticizing the hypocrisy and self-righteousness of the morally upright. Steve recognized his own self-righteous attitudes, and he started to realize that he, too, needed God's forgiveness.

He still had his doubts about the Christian faith, of course, but the more he talked with Drew and Matt and read the books they gave him, the smaller his doubts became. Problems that had initially seemed devastating to Christianity started to lose traction in his mind. And the arguments in favor of Christianity started to sound pretty compelling.

This was a welcome development to someone who, just a few months earlier, had struggled with obsessive thoughts about his own death. In Christianity, Steve found peace and hope of life beyond the grave. In a psychology class that semester, Steve had learned that religious people tend to be happier and better able to cope with death than people with no faith, and the more he embraced Christianity, the more he found it to be true for himself.

After half a year of exploration, Steve finally decided the time had come to make the leap of faith. Drew had given Steve a little booklet explaining Christianity, which included a short prayer at the end that someone could say if they wanted to "put Jesus on the throne of their life." Drew had emphasized that the words weren't some magical incantation—someone could obviously read them without truly meaning them, and the prayer wouldn't do anything. Similarly, someone could certainly become a Christian without saying a special prayer. "But if these words reflect the attitude of your heart," Drew had said, "then you should talk to God about it. And these are some words you can use to do that."

At one minute after midnight on April 12, 2009—Easter morning—Steve pulled out the booklet from his desk drawer and said the prayer alone in his dorm room.

It took him a few days before he said anything about it to Drew or Matt. He wasn't even sure how to bring it up: "Um, hey, I guess I'm a Christian now?" It felt sort of anticlimactic to just drop it into casual conversation, but when he finally did, his friends were ecstatic. Shortly after that, Drew encouraged Steve to get baptized.

And that's when Steve finally told me everything that had been happening in his life.

7

I SUPPOSE I SHOULDN'T HAVE BEEN surprised when Steve
tried to convert me. Not long after his baptism, he started
probing me with questions about my own religious views.

"Do you believe in God?"

"What do you think about the afterlife?"

"What do you think about Jesus?"

I quickly realized that I wasn't sure how to answer many
of his questions. I knew a decent amount about math and
science. I knew quite a bit about baseball. But I had thought
very little about God and the meaning of life.

I wasn't even sure what to call myself: An atheist? An
agnostic? To be honest, I hadn't had much reason to care
about the distinctions before now. But all of a sudden, with

Steve's pointed questions, I had to start thinking through what I actually believed about all these things.

One thing I knew for sure was that I believed in science. But I had never held the opinion that science disproves the existence of God. As the atheistic philosopher Bertrand Russell pointed out, there is no way for science to disprove the idea that the universe began five minutes ago with a built-in past. There is no way for science to prove that human beings have moral value. There is no way for science to prove that the laws of science that hold now will continue to hold a couple minutes from now. So there is no way that science could prove there isn't (or couldn't be) some sort of god behind all this.

But that didn't mean I was ready to embrace the God of conventional religion. I was intrigued by the idea of a deistic god, some impersonal force that set the world in motion and then dispassionately left it alone to run its course.

The big draw of deism was the fact that it didn't create any tension with science. There was no magic genie interfering with the universe—God just started the whole thing going and left the rest to the laws of nature.

But deism didn't do a very good job of explaining the fine-tuning of the universe for intelligent life. Even though I knew very little at the time about physics (and hadn't yet heard of string theory), I had somehow learned about fine-tuning: If the laws of our universe were changed even a little bit, life as we know it would be impossible. To many—including a young me—this concept pointed to some sort of a designer, a cosmic fine-tuner who set the knobs just where they needed to be to create a world with life. It was hard for me to believe that our world had arisen by pure chance, and it didn't make much sense that some impersonal force would want to create a world full of intelligent beings in the first place.

If God existed, I reasoned, he/she/it would probably be a tad more robust. I liked the idea, which I later learned had been famously advanced by evolutionary biologist Stephen Jay Gould, that certain questions belonged to science, and certain questions belonged to the realm of religion. Perhaps there was a God, responsible for things like creating the world, setting it in motion, and making the rules about right and wrong. Maybe science teaches us about how our universe works, how atoms move, and how the human body will respond to various chemicals, while religion teaches us about our meaning and purpose within that universe.

This sort of religion seemed reasonable enough to me, but atheism also seemed pretty appealing in its own right. I decided I was basically an agnostic: It would be nice if there were a God. It would be nice if there were some larger meaning and purpose. But I really didn't know for sure.

In the past, that had always been good enough: I figured it didn't really matter one way or the other, as long as I lived a decent and moral life. But now I was motivated to explore questions of faith for a very practical and selfish reason: I didn't want to lose Steve as my best friend. If I could find religious common ground with him without sacrificing my intellectual integrity, I would.

I told Steve I had no trouble believing in a god of some sort—largely due to the fine-tuning argument. As for Christianity, I was skeptical.

After I sent Steve a long message explaining my views on all of this, he replied, in part, "I definitely would have agreed with you not too long ago. But as I started to learn more, I found Christianity to be a lot more plausible than I expected it to be at the outset."

Steve passed on his copy of the book *Letters from a Skeptic*, which Matt had given him the year before. I was hesitant at first, and for a while the book sat untouched on the table next to my bed. But when I started to read it, I found it quite enjoyable.

The book consists of real-life correspondence between a Christian pastor named Greg Boyd and his skeptical father, Edward. First, Edward would write a letter explaining why he had trouble believing in God, and Greg would respond to Edward's objections. The book uniquely captures their interchange—real letters from real people wrestling with real questions about the existence of God. The tone was also important: Whereas many debates about the existence of God devolve into angry shouting matches, this debate took place against the backdrop of unconditional love between a father and his son.

As I read Edward's letters to Greg, I found myself resonating with just about everything he said. His objections to belief in God and Christianity were relatable, and he voiced them with more clarity than I could have done myself. Edward struck me as just a normal guy—raised in a culture like mine, with a respect for science and a hope that there might be something more to this world than meets the eye, yet with an intellectual recognition that these hopes were likely nothing more than wishful thinking.

"Why is the world so full of suffering?"

"Why does God create earthquakes and famines?"

"Why would an all-powerful God need prayer?"

"Isn't the Bible full of myths and God's vengeance?"

Every time Edward raised an objection in one of his letters, I found myself thinking that Greg would surely have a tough time answering it. But time and again, I was surprised by Greg's answers. They weren't always entirely convincing—for instance, I had a very hard time believing that all natural evil, such as earthquakes and famines, is caused by demonic forces, and his unorthodox view that God's omniscience is limited in some ways struck me as weird. But he clearly had thought a lot about all of these questions, and his answers were surprisingly not stupid. As I continued to read, I found myself becoming more and more convinced that Greg was onto something. And so was Edward: He converted to Christianity not long after the published correspondence ended, and he passed away shortly thereafter.

8

WHEN I FINISHED READING *Letters from a Skeptic*, Steve gave me another book about Christianity, called *I Don't Have Enough Faith to Be an Atheist* by Norm Geisler and Frank Turek. To be honest, I didn't like it quite as much as *Letters from a Skeptic*. Parts of it were pretty thorough and well-researched, but other parts were incomplete, overly polemical, and/or nonscientific. I had just completed my first college course on quantum mechanics, and I remember thinking that the book's presentation of the subject was grossly misleading.

That said, the book was very informative and offered a useful twelve-point framework for belief in Christianity.

I was quite happy with the book's first two points: "Truth about reality is knowable," and "The opposite of true is false." The more I learned, the more I was (reluctantly) willing to believe the third point: "The theistic God exists." But the fourth point is where the scientist in me started having trouble: "If God exists, then miracles are possible."

In my physics classes, I had been learning about the supposedly immutable laws of nature, and it bothered me to think that they might spontaneously break down. But it didn't take long for me to realize that "a spontaneous breakdown of the laws of nature" is not the right way to think about miracles. A miracle, I came to understand, is an act of divine intervention that doesn't *break* the laws of nature, but goes beyond their intended scope and domain of validity.

Somewhere in the process of my examination of Christianity, I came across a silly but useful analogy: Suppose I drop my phone. According to the law of gravity, the phone will fall downward until it hits the ground. But if I manage to reach out my hand in time, I can catch it before it lands. That doesn't mean I've violated the law of gravity. It's just that gravity isn't equipped to describe what happens if I intervene in the system.

Indeed, the laws of physics I learned in my introductory college courses are explicitly formulated to describe what will happen in a closed system that is free from outside intervention. The first law of thermodynamics says that the total amount of energy in a closed system remains constant over time; the second law says that the entropy of a closed system never decreases; Newton's first law of motion says that an object at rest will remain at rest in a closed system; and so on. Obviously, a system in which someone like God can intervene is not a closed system, so these laws would have nothing to say about that system. The laws of physics aren't *wrong*, but they are *incomplete*. Physics is not in the business of describing nature in the presence of supernatural forces.

Nonetheless, I rather liked the idea that our universe is a closed system in which the laws of nature apply universally. There might be some sort of lawgiver, some theistic God behind everything, but what evidence did I have that this sort of God would miraculously intervene in human affairs? When I posed this question to Steve, he turned it around: What evidence did I have that God *wouldn't* intervene?

It was true that I had never seen a miracle. But that didn't mean they didn't happen to other people. One day, Steve, Mom, and I were visiting my aunt in Florida and found ourselves discussing miracles at a restaurant.

"You know," my aunt said, "if I were to start speaking fluent Italian right now, everyone at this restaurant would assume I spoke Italian. Only the three of you would realize that something incredible was happening."

Truthfully, it wouldn't have been all that surprising to me if my aunt had picked up Italian at some point during her travels, but her point remained valid: Miracles might be happening all over the place but are only visible to a limited number of people. The fact that they have been hidden to me does not mean they are hidden to everyone. As the saying goes, "Absence of evidence is not (necessarily) evidence of absence."[3] Just because you haven't found something doesn't necessarily mean it isn't there. It could just mean that you're not looking in the right place.

So, where is the right place to look for a miracle? I wasn't sure, but I was quite certain that it wouldn't be in the realm of scientific experimentation. Sure, the entire field of chemistry indicates that water doesn't turn into wine, and the other miracles described in the Bible could not occur naturally. But the key word is *naturally*. Science can tell us quite a bit about the natural course of events, but it's not in the business of describing the supernatural.

Furthermore, the laws of science are defined by what people naturally and repeatedly observe. No scientist alive today has seen water become wine or a dead man rise from the grave. Hence, none of our scientific theories of chemistry or biology incorporate either of those phenomena (otherwise they could not properly be called miracles). In the end, I concluded that the apparent conflict between science and miracles was insufficient grounds for denying their existence; but it did mean I would have to search beyond the laboratory to find evidence of the miraculous.

Steve had apparently decided that the right place to find a miracle was in the Bible. I was skeptical. But I was willing to take a look.

9

"WHAT DO YOU WANT TO DO NOW?"

I must have posed this question to Steve more than a thousand times between the ages of five and nineteen.

His answer was usually predictable.

"Throw the football."

"Watch TV."

"Play video games."

"I don't know. What do you want to do?"

But one summer day in 2009, shortly after his baptism, he gave me a totally unexpected reply: "Let's do a Bible study."

"No thanks," I responded. "That's boring. Got any other ideas?"

In the end, I think we went out and played catch instead. Steve had stopped playing baseball after his sophomore year

in high school, but I continued to play—and train hard—
in the hope that one day my ability would take off. I took
round after round of batting practice and threw bullpen after
bullpen—never seeing much improvement, but never losing
hope that my skills would improve if I just worked a little
bit harder.

Of course, I needed a partner to catch me and throw batting practice, and that job naturally fell to Steve. Though his
own baseball career was over and he had no personal reason to practice, he was always willing to help me out. He
would throw me a hundred pitches of batting practice and
take maybe twenty swings himself. He hit ground ball after
ground ball so I could practice my fielding, without asking
for any himself. And in the summer of 2009, as I desperately
tried to resuscitate a baseball career that was clearly over by
learning how to throw a slider, Steve sat behind the plate,
without catcher's pads, while I buried pitch after pitch in the
dirt in front of him, often taking balls to the shins, chest, and
face in the process.

When I told Steve I didn't want to do a Bible study, he
didn't argue with me. He just kept quiet, let it go, and did
whatever I wanted to do instead. At first I was fine with that,
but my conscience wouldn't let it go. For the next few hours,
I thought back to all the rounds of batting practice, all the
ground balls, all the pitches in the dirt. Steve had probably
spent a hundred hours doing something he had no use for just
because it would make me happy. And now he wanted just
thirty minutes to do something for which I had no use—read
the Bible together. I decided I should probably do that for him.

"Hey Steve, do you still want to do that Bible study?"

"Sure."

We started by reading the first eighteen verses of the Gospel of John together:

In the beginning was the Word, and the Word was
with God, and the Word was God. He was with God
in the beginning.

Through him all things were made; without him
nothing was made that has been made. In him was
life, and that life was the light of men. The light
shines in the darkness, but the darkness has not
understood it.

There came a man who was sent from God;
his name was John. He came as a witness to testify
concerning that light, so that through him all men
might believe. He himself was not the light; he came
only as a witness to the light. The true light that
gives light to every man was coming into the world.

He was in the world, and though the world was
made through him, the world did not recognize
him. He came to that which was his own, but his
own did not receive him. Yet to all who received
him, to those who believed in his name, he gave the
right to become children of God—children born
not of natural descent, nor of human decision or a
husband's will, but born of God.

The Word became flesh and made his dwelling
among us. We have seen his glory, the glory of the

One and Only, who came from the Father, full of
grace and truth.

John testifies concerning him. He cries out,
saying, "This was he of whom I said, 'He who comes
after me has surpassed me because he was before
me.'" From the fullness of his grace we have all
received one blessing after another. For the law was
given through Moses; grace and truth came through
Jesus Christ. No one has ever seen God, but God the
One and Only, who is at the Father's side, has made
him known.[4]

"Any observations?" Steve asked.

"I guess 'the Word' here is talking about Jesus?"

"Yes, that's right." Steve explained to me the remarkable,
paradoxical claims that John was making about Jesus. He is
both *with* God and actually *is* God. He is both the creator of
humankind and also has made his dwelling on earth, among
humanity. We talked about the meaning of the phrase, "the
darkness has not understood it," and Steve pointed out that
some translations read, "the darkness has not overcome it."
We agreed that the latter translation was easier to understand. At one point, Steve remarked that the John referred
to in the passage is not John the author, but rather John the
Baptist.

"Whenever John the author wants to refer to himself, he
uses the phrase—"

"The one whom Jesus loved," I interrupted.

"Yeah, how did you know that?" Steve was impressed

THE ROAD TO FAITH ≈ 55

that I knew anything about the Bible at all. I had somehow remembered that fact about John from one of the books on Christianity I had read recently. When our study ended, I felt a sense of accomplishment, knowing I had made Steve proud of me. In retrospect, it was pretty vain—I was still focused primarily on my own reputation, so my assessment of the Bible study was based more on how I had performed in front of Steve than on what I had learned about anything. Nevertheless, our time together showed me that a Bible study didn't have to be boring, strange, or awkward. It was just my brother and me discussing an interesting book the way we might have discussed a good article in *Sports Illustrated* or *ESPN: The Magazine*.

10

"**LOOK, STEVE, I HAVE TROUBLE** finding time to read books I *want* to read, much less books I *don't* want to read."

That's what I told my brother the first time he tried to give me a copy of the Bible. Though I had never read it, I just knew it had to be boring. There's even an episode of *The Simpsons* in which Homer, Marge, Bart, and Lisa all fall asleep in church due to the heat and boredom. The rest of the episode consists of their respective dreams—essentially Bible stories in which the biblical characters are replaced by characters from the show. In Bart's dream, for example, he is David, and Nelson, the school bully, is the giant Goliath.

The Simpsons' dreams were filled with humor, so they were entertaining to watch. But the Bible itself? It seemed as if all my friends who attended church growing up went mainly

because their parents made them, and they couldn't wait for it
to be over. Conversations with them had led me to believe that
the Bible was a boring list of dos and don'ts—commandments
that were either common sense ("thou shalt not murder") or
else hopelessly strict ("thou shalt not take the name of the Lord
in vain")—interspersed with folktales designed to teach moral
lessons, on a par with Aesop's Fables or Dad's fox stories.

Eventually, however, I accepted a copy of the New
Testament from Steve and actually started to read it. I did
this for two reasons: For one, I didn't want my twin brother
to look down on me for refusing a challenge. In my mind,
Steve was now a Bible freak, with verses on sticky notes hanging over his desk, and I hadn't even cracked the cover. The
second reason was simple curiosity: The Bible he gave me
wasn't like any I had seen before. For one thing, it was a
copy of the New Testament only, which meant it was about
a quarter of the thickness of a full Bible, and thus far less
intimidating. On top of that, the version Steve gave me was
written in modern prose, which for me was like a breath of
fresh air. In my ignorance, I had assumed that the Bible was
written in Old English, full of *thees* and *thous*. I was pleasantly surprised to learn that the original New Testament was
actually written in Greek—and not just any Greek, but *Koine*
Greek, the common form that even a kid could understand.

As soon as I started reading, I realized that my preconcep-
tions were off base. I had committed to reading one chapter
of the Bible every day, but there were days when I couldn't
put the book down. Whatever else the Bible might be—
strange, inaccurate, far-fetched—it certainly wasn't boring.

11

I STARTED READING AT THE BEGINNING, the book of Matthew, which opens with a genealogy of Jesus. This didn't make much sense to me at the time. Chapter 2 features the classic Christmas story, in which the three wise men bring gifts of gold, frankincense, and myrrh to the baby Jesus. This was a story that even I was familiar with.

After reading stories about John the Baptist and about Jesus being tempted by Satan in the wilderness and then gathering some followers, I got to the celebrated Sermon on the Mount, a collection of Jesus' moral teachings. There I found the Golden Rule, which I had been taught since my youth: "Do to others whatever you would like them to do to you" (Matthew 7:12), as well as a number of other dos and don'ts. The rules seemed harsh at times—as I expected from a religious text—but at the same time they offered a surprisingly compelling vision of the way things should be: "You have heard the law that says, 'Love your neighbor' and hate your enemy. But I say, love your enemies! Pray for those who persecute you!" (Matthew 5:43-44).

One thing I started noticing about Jesus was that he wasn't above associating with the poor, the meek, and the outcasts, even when the cool kids mocked him for it:

> Matthew invited Jesus and his disciples to his home as dinner guests, along with many tax collectors and other disreputable sinners. But when the Pharisees saw this, they asked his disciples, "Why does your teacher eat with such scum?"
>
> When Jesus heard this, he said, "Healthy people don't need a doctor—sick people do." Then he added, "Now go and learn the meaning of this Scripture: 'I want you to show mercy, not offer sacrifices.' For I have come to call not those who think they are righteous, but those who know they are sinners."[5]

At a time when disability was seen as an indication of moral poverty and sin (see John 9:2-3), Jesus rebelled against the status quo and made a special point of caring for the lame, the blind, and the deaf:

Jesus returned to the Sea of Galilee and climbed
a hill and sat down. A vast crowd brought to him
people who were lame, blind, crippled, those who
couldn't speak, and many others. They laid them
before Jesus, and he healed them all. The crowd was
amazed! Those who hadn't been able to speak were
talking, the crippled were made well, the lame were
walking, and the blind could see again! And they
praised the God of Israel.[6]

There seemed to be no limit to how far Jesus would go
for the cause of the weak. In a culture where touching a
leper made one ceremonially unclean, Jesus touched the lepers *before* he healed them:

Large crowds followed Jesus as he came down
the mountainside. Suddenly, a man with leprosy
approached him and knelt before him. "Lord," the
man said, "if you are willing, you can heal me and
make me clean."
　　Jesus reached out and touched him. "I am
willing," he said. "Be healed!" And instantly the
leprosy disappeared.[7]

These stories about Jesus reminded me of third grade,
when I sat next to a girl named Amber, who wasn't very
popular. I was the new kid in town and wasn't very popular
either, but my athletic ability at least gave me an in with the

jocks, who would pick me for their teams at recess. This left me with a choice about Amber: Either befriend her and risk whatever "coolness" being a jock had earned me, or ignore her like everyone else did.

As a senior in high school, I sat next to a girl in AP Spanish named Marta (at least, that was her Spanish name), who also wasn't very popular. She was kind of eccentric and somewhat of a teacher's pet, and thus she was scorned by the other students. This left me with a choice about Marta: Either befriend her or reject her like everyone else did.

Both times, I chose the latter option: I rejected the outcasts rather than join in their suffering. From a cost-benefit perspective, it only made sense: If I befriended people with no social capital, I would gain nothing. In fact, I would probably lose what little social capital I could gain on my own because I would stand out from the crowd. If you want to move up in the world, it doesn't make sense to mess around with the people below you. You need to cozy up to the people at the top and hope that their acceptance will accrue to your account.

This was not at all how Jesus went about things. He was full of compassion and didn't care what the so-called people at the top thought about him. Whereas I wasn't willing to befriend someone if it meant being picked last at recess, Jesus laid his reputation, his health, even his life on the line for the sake of people in need.

Equally impressive, yet often overlooked, was the fact that Jesus wasn't always Mr. Nice Guy. Whereas so much of my energy was spent trying to impress the popular guys and cute girls, Jesus was totally unconcerned with his reputation and didn't mind sticking it to people when he had to:

Woe to you, teachers of the law and Pharisees, you hypocrites! You shut the door of the kingdom of heaven in people's faces. You yourselves do not enter, nor will you let those enter who are trying to.

Woe to you, teachers of the law and Pharisees, you hypocrites! You travel over land and sea to win a single convert, and when you have succeeded, you make them twice as much a child of hell as you are.[8]

When people with selfish ulterior motives came to Jesus seeking to condemn him and justify themselves, Jesus responded without apology:

Some Pharisees and teachers of religious law now arrived from Jerusalem to see Jesus. They asked him, "Why do your disciples disobey our age-old tradition? For they ignore our tradition of ceremonial hand washing before they eat."
Jesus replied, "And why do you, by your traditions, violate the direct commandments of God? For instance, God says, 'Honor your father and mother,' and 'Anyone who speaks disrespectfully of father or mother must be put to death.' But you say it is all right for people to say to their parents, 'Sorry, I can't help you. For I have vowed to give to God

what I would have given to you.' In this way, you
say they don't need to honor their parents. And so
you cancel the word of God for the sake of your own
tradition. You hypocrites! Isaiah was right when he
prophesied about you, for he wrote,

'These people honor me with their lips,
      but their hearts are far from me;
   Their worship is a farce,
      for they teach man-made ideas as commands
         from God.'"[9]

Jesus' disciples came to him in private after this and asked
whether he realized he had offended the Pharisees with his
teaching. Ha! As if Jesus gave a rip about what the powers that be thought of him! In the same breath, Jesus could
speak tender love to the "sinners" and spit fiery truth to the
self-righteous religious elites, entirely unconcerned about any
personal cost to himself.

And yet it was clear that he didn't stand up to those in
power because he hated them; he did it because no one else
would tell them what they needed to hear. Indeed, Jesus
expressed remorse over the lost religious leaders just as he
did over the lost sinners:

O Jerusalem, Jerusalem, the city that kills the
prophets and stones God's messengers! How often
I have wanted to gather your children together as a

hen protects her chicks beneath her wings, but you wouldn't let me.[10]

To this day, one of the biggest regrets of my life is the way I treated Amber and Marta. I realized that I was a Pharisee, loving the place of honor at the feasts yet neglecting the weightier matters of justice, mercy, and faithfulness. But Jesus was different. The radical compassion, the self-jeopardizing opposition to the powerful, the deep love for all humanity was not just a fantasy; it was a prophecy that Jesus himself was fulfilling even as he proclaimed it.

And that was from just the first book of the New Testament.

12

I WAS (AND STILL AM) A BIG FAN of grunge and alternative rock music. Steve prefers music that isn't quite as loud and angsty, but there's enough overlap in our tastes that we can generally settle on a radio station. If not, whoever's driving has priority.

The summer of our first Bible study, on the many trips we took between our parents' and grandparents' houses, Steve and I heard plenty of each side—Jack Johnson, the Goo Goo Dolls, and John Mayer when Steve was driving, and the Foo Fighters, Pearl Jam, and Red Hot Chili Peppers when I was behind the wheel.

One night as I was pulling the car out of our grandparents' garage, one of my favorite songs—"Lightning Crashes" by the band Live—came on, and I wanted Steve to appreciate it as much as I did.

"This is a great song," I said. "Listen to the lyrics." We sat in silence and listened. The song tells of some sort of life force that departs from a dying woman in a hospital room but comes back to a newborn baby down the hall. It's a story about the circle of life, the hope that the dead are not really gone forever, but that their life force will live on in an eternal cycle of death and rebirth.

I thought Steve would like the song because it was essentially a religious message, and Steve was now a religious person. Once again, it was silly of me to lump all religions into one category. Clearly, the Eastern view of reincarnation hinted at by the song is very different from the Christian view of bodily resurrection, but I didn't understand that at the time. So I was surprised when Steve didn't like the song. Instead, he gave its view of the afterlife a rather condescending, "Ha, that's weird."

This angered me, not because I personally subscribed to the song's spiritual views, but simply because I liked the music and didn't like having my musical tastes dismissed. I bit my tongue, but I wondered, *What makes Steve think his spiritual views are superior to these?*

Why Christianity, as opposed to any other religion? Practically speaking, Christianity was the sensible place to start my spiritual journey. My family wasn't very religious, but we did celebrate the Christian holidays of Christmas and Easter. The fact that Steve had become a Christian meant I would have a brother to guide me in my search. And if I did become a Christian, Steve and I could bond over a common faith, rather than arguing about our differing views of religion.

But ultimately, of course, I didn't want to convert to Christianity solely for practical reasons. If there was a better religion out there—Islam, Judaism, Buddhism, Voodoo, Deism, New Age spirituality—I would like to have found it.

Yet, as I began to explore the different faiths, I quickly found Christianity more appealing than the others. I was attracted by the fact that Christianity claims to offer good news, not merely good advice. I have since read the entire Hebrew Bible and the English translation of the Quran. Both have a large number of commandments, laws, and suggestions for how to live fruitful lives. And for the most part, I find them both to be quite reasonable. Buddhism, too, is full of good advice, and in recent years my spiritual and mental health have benefited greatly from Buddhist meditation principles and techniques.

But if all I wanted was advice on how to live a happy life, there were other places to turn besides religion—my parents, my teachers, the textbook from my introductory psychology class, and the internet all had plenty to say. But Christianity promised something more than just good advice: It promised eternal life.

Eternal life wasn't merely something I had to take on blind faith, just because a wise man said so. After all, my middle school math teacher was a very wise man, but that doesn't mean I would trust him if he tried to sell me on a new religion he had just invented. If I were going to believe any religious leader, I wanted some justification as to why I should trust whatever he or she said. Christianity at least claimed to offer such a justification—namely, the resurrection of Jesus. In other words, the Christian faith isn't founded on purely religious or metaphysical thinking; it is founded on a historical person and a historical event.

That's not to say I believed in the resurrection of Jesus at the time—it took a while before I became convinced of that. But it was attractive, at least, that Christians were not merely claiming that Jesus had risen from the dead into some make-believe land of the gods. They said he had actually lived, died, and risen from the dead into our space-time universe as an actual event in history. And that meant I could investigate the claims of Christianity, just as I would any other event of history. As such, Christianity offered something especially appealing to the budding scientist in me: a falsifiable prediction, and a testable basis for its metaphysical claims.

Most of all, Christianity offered Jesus. Throughout history, so many people have claimed divine authority and used it to accumulate wealth, power, and women, or to justify their evil deeds. Hitler believed he was on a divine mission to rid the world of Jews, homosexuals, and the Roma people; in fact, the Nazis' belt buckles were inscribed with the words *Gott Mit Uns*—"God with us." Pope Urban II claimed it was God's will that the Christians launch a bloody crusade to reclaim the Middle East from the Muslims. The 9/11 terrorists thought they were waging a divinely inspired holy war against an evil nation. And L. Ron Hubbard, the founder of

Scientology, is reported to have said on multiple occasions, "If you want to get rich, start a religion."

You look at those guys, and then you look at Jesus. Jesus also claimed divine authority, but what did he do with it? What was his great plan? He loved those who hated him, he shared meals with sinners, he challenged those in power, and he rose to prominence not through military might, but through self-sacrificial love.

I was struck by a sermon Steve shared with me called "One Solitary Life" by James Allan Francis:

> He was born in an obscure village, the child of a
> peasant woman. He grew up in still another village,
> where He worked in a carpenter's shop until He
> was thirty. Then for three years, He was an itinerant
> preacher. He never wrote a book, never held an
> office, never had a family or owned a house. He
> never went to college. He never visited a big city.
> He never traveled two hundred miles from the place
> where He was born. He did none of the things
> one usually associates with greatness. He had no
> credentials but Himself.
>
> He was only thirty-three when the tide of public
> opinion turned against Him. His friends ran away.
> He was turned over to His enemies and went
> through the mockery of a trial. He was nailed to
> a cross between two thieves. While He was dying,
> His executioners gambled for His clothing, the only
> property He had on earth. When He was dead, He

was laid in a borrowed grave through the pity of
a friend.

Twenty centuries have come and gone, and
today Jesus is the central figure of the human race
and the leader of mankind's progress. All the armies
that ever marched, all the navies that ever sailed,
all the parliaments that ever sat, all the kings that
ever reigned put together have not affected the life
of mankind on this earth as much as that ONE
SOLITARY LIFE.[11]

If nothing else, the unparalleled popularity of Christianity
worldwide was a sign that I wasn't totally unreasonable for
thinking there was something special about Jesus. Christianity
has become the most popular religion in the world, tran-
scending racial, social, and cultural lines, which (a) is what
I would have expected of a true religion, if such a thing
were to exist, and (b) is a testament to the uniqueness of the
Christian drama and the character of Jesus. Doubtless, some
of the spread of Christianity has involved violence, and world
missions have often become entangled with the evils of western colonialism. But in the beginning, Christianity spread
like wildfire despite persecution from the Roman Empire.
Even today, Christianity is seeing rapid growth in Africa and
Asia, driven primarily by the efforts of local churches (along
with higher birth rates) rather than foreign missionaries.

The world seemed to really like Christianity. Maybe they
were onto something.

13

MY SEVENTH GRADE BIOLOGY TEACHER, Ms. Hankins, once gave the class a five-minute pep talk on evolution. She explained to us that "creation science," which denies that humans evolved, is not actually science—it is pseudoscience, and none of us should believe it.

Now, five minutes is not a very long time. Perhaps if our classes had lasted longer than forty-five minutes, Ms. Hankins would have gone into more depth on the subject and given us a real argument as to why evolution is the correct explanation for the diversity of life, rather than merely asserting it as a fact. If there had been any hard-core young-earth creationists in that class, I'm sure they would have been thoroughly unconvinced by anything she told us that day.

And yet we all trusted Ms. Hankins, even without actually poring over the evidence ourselves. Why? Because we already had a good reason to believe her—she was an excellent teacher who knew her subject well. Over the course of the year, she had established herself as a trustworthy source.

Trustworthy sources are often hard to find, but they're very important. Sure, there are people like Srinivasa Ramanujan—the Indian mathematician who grew up without access to textbooks or proper mathematical instruction yet managed to single-handedly reproduce dozens of major results in number theory and discover many more of his own. But for the rest of us, reliance on the work of others is crucial. I never could have invented calculus by myself, yet thanks to Isaac Newton's insights and my high school math teachers' instruction, I now find it to be very easy.

When it came to assessing the historical truth of Christianity, I found myself struggling with the question about whether I could trust the sources presented to me. There is a widespread belief that our only knowledge of the life of Jesus comes from the Bible, and this just isn't true. There are more than a dozen non-Christian sources from the first couple of centuries AD that mention Jesus, several of which even mention his death by crucifixion. But it is true that a large portion of what we know about Jesus comes from Christian sources—the Bible especially. And this raises the crucial question: Are these sources trustworthy?

The answer you get to this question depends on who you ask. Christians will point to a number of reasons why the Gospel accounts of Jesus' life should be considered more trustworthy than other ancient biographies, whose trust-worthiness is largely taken for granted. For example, the Gospels were written much closer to the time of the events they describe than just about any other ancient histories; there are far more copies of them than other ancient sources; there are very early copies of them; and there are multiple Gospels, offering multiple testimonies on the life, death, and resurrection of Jesus.

But others are more skeptical. They will argue that the Gospels were written decades after the events they describe, most likely by people who weren't even eyewitnesses, and all four carry a strong pro-Christian bias. Who knows if what they say about Jesus is true?

I, for one, certainly didn't feel qualified to determine who was right here: the Christians or the skeptics. If I was going to believe in Christianity, I thought, I would need to find a source whose trustworthiness is more certain. I found it in the apostle Paul.

Paul is said to have written thirteen of the twenty-seven books in the New Testament. Scholars today—especially the skeptical ones—attribute only seven of these confidently to Paul, and they are on the fence with a couple of others. But even when I looked only at those seven—Romans, 1 Corinthians, 2 Corinthians, Galatians, Philippians, 1 Thessalonians, and Philemon—I got a heavy enough dose of Paul to realize he had an amazing story to tell.

Paul was not always a follower of Jesus. In fact, he was once a passionate enemy of Christianity. Paul himself writes, "For you have heard of my previous way of life in Judaism, how intensely I persecuted the church of God and tried to destroy it. I was advancing in Judaism beyond many of my own age among my people and was extremely zealous for the traditions of my fathers."[12] According to the New Testament book of Acts, Paul was on his way to Damascus to persecute Christians when he had an experience of the resurrected Jesus, an appearance which is further attested in Paul's first letter to the Corinthians. As a result, Paul immediately changed teams and joined the Christians he had been persecuting just shortly before.

When I first read this, it struck me what an incredible life change this must have been. Paul had spent his entire life believing that God was totally above and beyond any material thing in this world. Whereas pagans (and disobedient Jews) would worship statues of other gods, Paul was a devout Jew and never would have given his allegiance to anything but the Jewish God. He was so committed to his worldview that he went around overseeing the executions of Christians.

Then, all of sudden, he reversed his entire purpose and mission in life and started helping the people he had been trying to eradicate. Paul subsequently wrote that Jesus was "in very nature God."[13] So much for not worshiping any material thing!

Paul went on to live a life totally devoted to making Jesus' life and resurrection famous. In one of his letters, Paul even boasts about all of the torture he underwent for the sake of Jesus:

Five times I received at the hands of the Jews the
forty lashes less one. Three times I was beaten
with rods. Once I was stoned. Three times I was
shipwrecked; a night and a day I was adrift at sea;
on frequent journeys, in danger from rivers, danger
from robbers, danger from my own people, danger
from Gentiles, danger in the city, danger in the
wilderness, danger at sea, danger from false brothers;
in toil and hardship, through many a sleepless night,
in hunger and thirst, often without food, in cold
and exposure.[14]

These are not the words of a liar. Clearly Paul sincerely
believed his experience of the risen Jesus was an authentic,
miraculous revelation from God himself, which had completely changed his worldview. And for someone in my
position, Paul was what I needed to get started in my investigation of the historical truth of the Resurrection. I soon came
across 1 Corinthians 15, Paul's list of Jesus' post-resurrection
appearances, considered by scholars across the board to be
the most significant testimony on Jesus' resurrection. From
there, I learned about James, the brother of Jesus, who was
another nonbeliever converted to Christianity by an encounter with the resurrected Jesus, and who also ultimately died
a martyr's death for his faith.

Especially in my early days of researching the historical
evidence for Christianity, Paul offered a testimony that I
couldn't easily ignore, even if I had wanted to. I found him

relatable—he was so firmly entrenched in his previous ways of thinking that he never would have changed them without tangible, solid evidence to the contrary. But in other ways, Paul's life was very different from mine: He had given up everything to follow Christ. Reading Paul's letters, I got a glimpse into what God expected of me as a Christian—unconditional surrender, uncompromising faithfulness.

I wasn't ready to make that kind of commitment to God, even if he did exist.

14

IN HIGH SCHOOL, WE SOMETIMES had to read multiple articles of opposing viewpoints whenever there was a controversial issue. And time after time, I would read one side of the story and agree with it, but then read the other side and agree with it too. After all, the world is full of shades of gray, and things are rarely as cut-and-dried as they seem at first glance.

As I read the New Testament and went to church with Steve, Christianity began to resonate with me. Most of the stereotypes I'd held had been addressed: The Bible wasn't boring. Reasonable people believed it. Miracles didn't contradict science—in fact, the fine-tuning of the universe for intelligent life might even point to some sort of cosmic designer. But maybe this was just another one of those times where I was convinced because I had seen only one side of the story. Perhaps once I looked at the other side, I would be equally convinced that atheism was the more reasonable position.

Lucky for me, Steve had also brought home Richard Dawkins's *The God Delusion*, one of the most famous atheistic books of the early 2000s. Steve had already seen what can happen when people close themselves off to arguments from the other side. Every year, Christians show up as college freshmen who have never had their faith tested. Maybe their youth pastor back home gave them the mantra "Don't ask questions, just have faith." Then they go to college, and all of a sudden, they're faced with intellectual and social pressures they've never seen before: intelligent skeptics ask them probing questions, professors poke fun at the whole concept of religion, their roommates invite them to put down that silly Bible and come have fun, and pretty soon they've abandoned their faith entirely.

But "just have faith" wasn't good enough for Steve, or for me. I wanted to know what the objections to faith were, and I wanted to have those objections answered, if they could be. So I started to read *The God Delusion*.

To my surprise, I didn't find Dawkins at all convincing. Much of the book is dedicated to ranting about the stupidity of young-earth creationists. This seemed unnecessary: I wasn't a young-earth creationist and wasn't planning to become one any time soon.

When *The God Delusion* took a break from its attack on

creationism to address the tenets of Christianity, it was very unpersuasive. I almost stopped reading entirely when I got to what Dawkins calls the "central argument of my book," which goes roughly like this:

1. It has been a great challenge to explain why the universe looks designed.
2. "The natural temptation is to attribute the appearance of design to actual design itself," as we do with most things that have been designed (e.g., a watch).
3. "This temptation is a false one, because . . . [it] raises the larger problem of who designed the designer." In other words, the designer solution is even more improbable and complicated than the problem it was meant to solve.
4. Darwinian evolution has shown that the apparent design in biology is just "an illusion."
5. "We don't yet have an equivalent" mechanism to explain the apparent design manifest in the laws of physics, though "some kind of multiverse theory could in principle do [the job]."
6. "We should not give up hope of a better crane [i.e., a better mechanism to explain the apparent design] arising in physics, something as powerful as Darwinian evolution is for biology. But even in the absence of a strongly satisfying crane to match the biological one, the relatively weak cranes we have at present are, when abetted by the anthropic principle,

self-evidently better than the self-defeating skyhook
hypothesis of an intelligent designer."

7. Therefore, "God almost certainly does not exist."[15]

Even without any formal training in philosophy, I spotted
some huge holes in these arguments. First off, the conclusion
that "God almost certainly does not exist" doesn't follow from
the first six points of the argument. At best, his argument
would show that the apparent fine-tuning of the universe for
intelligent life is not a good argument for God's existence. But
just because one argument for God's existence fails, it doesn't
mean that *every* argument fails. Perhaps one of the other arguments would work better? To quote the atheistic philosopher
Kai Nielsen, "To show that an argument is invalid or unsound
is not to show that the *conclusion of the argument is false*. . . .
All the proofs of God's existence *may* fail, but it still may be
the case that God exists."[16] Even if *The God Delusion* had
managed to refute every single argument for the existence of
God, it still wouldn't show that God doesn't exist.

But Dawkins wasn't even claiming to address every argument here—he was focusing on one particular argument for
the existence of God and claiming it was rubbish. But he
didn't even do a very good job on that front. Dawkins asks
the question, "If God designed the universe, what designed
God?" Theistic philosophers would surely have better answers
to that question than I did, but even with a rudimentary
knowledge of philosophy, I could see that the demands
Dawkins was making of theism were far too stringent—few
theories of science could stand up to such scrutiny. String

theory, for instance, is far and away the best attempt to reconcile the two most important paradigms of modern physics: quantum mechanics and Einstein's general relativity. And yet, string theory is vastly more complicated and more difficult to explain than either quantum mechanics or Einstein's general relativity. But that doesn't give us a reason to reject it.

Atheistic philosopher Peter Lipton gives even more examples in a discussion about understanding *explanation*: "Explanations need not themselves be understood. A drought may explain a poor crop, even if we don't understand why there was a drought; I understand why you didn't come to the party if you tell me you had a bad headache, even if I have no idea why you had a headache; the big bang explains the background radiation, even if the big bang is itself inexplicable, and so on."[17]

So the central argument of *The God Delusion* wasn't very compelling to me. In fact, it was even worse than that: The book drew attention to the fact that the atheistic worldview into which I had slowly been diffusing actually required faith in its own right. Look again at premises 5 and 6 of Dawkins's argument:

5. "We don't yet have an equivalent" mechanism to explain the apparent design manifest in the laws of physics, though "some kind of multiverse theory could in principle do [the job]."
6. "We should not give up hope of a better crane arising in physics, something as powerful as Darwinian evolution is for biology."[18]

These premises are almost indistinguishable from the statement "We might be able to find more evidence against the existence of God, and then we'd have more evidence against the existence of God. So let's not give up hope that someday we will have more evidence against the existence of God."

For most of my young life, I realized, I had implicitly taken naturalism to be the null hypothesis—that is, in the absence of strong evidence to the contrary, I should assume that this universe is all that exists. But the more I learned, the more I started to recognize that atheism was not immune to criticism. Atheism is not merely a denial of certain doctrines and philosophical axioms, but instead offers its *own* set of doctrines and philosophical axioms about the ultimate nature of reality. The hypothesis that our extraordinary universe came into existence "from nothing, by nothing, and for nothing"[19] may ultimately be correct, I reasoned, but it wasn't worthy of being the null hypothesis.

I still wasn't ready to make the leap of faith, but I was much closer than when I had started.

Part II

No Such Agency

15

THE NATIONAL SECURITY AGENCY, or NSA, is part of the US
intelligence community. It is sometimes referred to jokingly
as No Such Agency, due to its high levels of secrecy. It is also
the largest employer of PhD mathematicians in America and
likes to recruit college students for summer internships in the
hopes of bringing them on later for full-time positions. In
2010, I was one of those college students.

When I first received an internship offer from the NSA,
I was ecstatic. It was clear by this point that I wasn't going
to cut it as a professional baseball player, but this internship
could put me on the road to the next best thing: working for
a federal intelligence agency. The Bourne movies had become
some of my favorites by this time, and imagining myself as

a combination of James Bond and Jason Bourne resulted in an unfettered optimism about my prospective internship. Of course, my position would involve far fewer guns and beautiful women and a lot more math.

Once the mathematicians at the NSA decide that your credentials qualify you for an internship, they request a top secret clearance for you from their friends in the security department. Before you can be granted such a clearance, you must (a) clear a background investigation, (b) complete a psychological evaluation, and (c) pass a polygraph test.

The background investigation itself is startlingly thorough. I had to provide a reference for every place I had worked or lived in the past ten years, which was a lot given how much my family had moved around. Then they actually sent agents, in person, to visit these references. In some cases, they even asked my references for more references without notifying me. All of a sudden, I was getting texts from my college friends saying they had just been contacted by a federal investigator regarding me and asking what was going on.

Not even the slightest hint of possible trouble went undocumented. One of my college roommates held a dual citizenship with the Philippines, a fact which drew a handful of questions to ensure he wasn't a Filipino spy. They were also very curious about any sexual relationships I might have been in (since these can be exploited via blackmail to gain top secret information). Fortunately for me, I didn't have anything to talk about.

I passed the background investigation without difficulty. My friend Nick told them I was "pretty much the perfect guy." My other friends attested similarly to the quality of my character. When they asked one of my coworkers in the intramural sports department at Cornell whether I was engaged in any acts of terrorism, he responded, "Are you kidding? We work together as referees. Terrorist plots are not one of our main topics of conversation."

Pretty much the only blemish on my record was the fact that I, like virtually every other kid I knew, had illegally downloaded music through an online music-sharing website. Throughout the course of my security clearance process with the NSA, I had to talk with six or seven different people about that measly fifty bucks' worth of illegally downloaded music.

Nevertheless, my references had spoken well of me, and now it was time for the NSA to evaluate me in person. They flew me down to Linthicum, Maryland, for a couple of days and put me up in a hotel. The first day would include a psychological assessment test and several interviews with HR, a psychologist, and operations. The itinerary for the second day showed only a polygraph at 8:00 a.m. and lunch at noon.

"Four hours for a polygraph?" I asked when I received my itinerary. The woman who had sent me the schedule responded that they don't like to schedule anything else on the day of the polygraph, firstly because it takes so long, and secondly because it is "rather stressful."

I found this highly amusing. Pitching against my twin brother in high school was stressful. My first physics prelim was stressful. Answering a bunch of questions about my squeaky-clean lifestyle? This was going to be a walk in the park.

16

THE NSA FACILITY LOOKED PRETTY MUCH like you would expect a top secret government facility—or a prison—to look. Barbed wire fences, guards with guns, security badges on everyone. I got a security badge with a big "A" for "applicant."

I was joined by a handful of others from all walks of life. Some were in their forties and fifties. Some looked like college students or recent college grads. One unfortunate high school internship applicant hadn't brought his passport or driver's license but only a school ID, and he got chewed out big-time by security.

"You are applying for a top secret security clearance! This is not an acceptable form of identification!" I'm pretty sure they made him leave without an interview that day.

For Christmas, Mom had given me a metal briefcase like the ones used in action movies for carrying guns and money. "What have you got in there, a machine gun?" the lady monitoring the metal detector joked. She let me through after checking that its true contents were much less interesting.

The psych assessment consisted of several hundred questions designed to ensure I wasn't a psychopath. It was straightforward for the most part, but one question gave me pause: "To what extent do you agree with the following statement: *Dying would be a relief*?"

The choices were "Completely true," "Mainly true," "Somewhat true," or "Not true." My mind went immediately to the Bible verse I had recently read: "To live is Christ and to die is gain."[1] On the other hand, I honestly wasn't too keen on dying, and I definitely did not want to appear suicidal. "Somewhat true," I answered.

I don't even remember the HR interview, but I'm sure it was the usual, boring, administrative stuff. So the next important event in my day was my interview with the psychologist. Amazingly, he had actually read through all my answers on the psych assessment battery, and he immediately brought up my answer to the "dying would be a relief" question. I suppose even "somewhat true" was enough to raise the possibility in his mind that I was thinking of killing myself.

"No," I said, "I'm a Christian. That answer was based on my religious beliefs."

Did I really just say that? I know I wanted it to be true—I wanted to call myself a Christian—but as the words came out, I felt like a phony. When I looked at Steve, I could see

a significant, tangible change from the person he had been
a year ago. But me? Had my life even been affected at all?
Sure, I had learned enough to believe that Christianity was
a reasonable worldview—perhaps even the *most* reasonable
worldview. But was it really *my* worldview?

Nonetheless, I had for the first time identified myself as
a Christian, and not just because I wanted to fit in with my
friends or pretend I was "holy." I had stood up for my faith,
knowing that the psychologist could have labeled me a suicidal religious zealot and ended the whole process right there.
The fact that he was satisfied by my answer was the first bright
spot of my day. Unfortunately, it would also be the last.

The operational interviews were conducted in pairs, and
I was matched with another applicant from the University
of Illinois Urbana-Champaign. The first thing they told us
was that we already had the job, and we weren't about to lose
it—unless of course we couldn't manage to earn our security
clearances. They really just wanted to see how we could function as a team.

They gave us an unsolved math problem and had us work
on it for fifteen minutes while they observed our teamwork
and progress.

When I'd heard my partner was from the University of
Illinois, I pitied his lowly, state school education. I, the Ivy
League Man, was certainly going to need to do most of the
work.

Nope.

I have never been so thoroughly embarrassed intellectually in my life. It turned out we were a one-man team, with

my partner figuring out more in the first minute than I was able to deduce in the entire fifteen. In the end, they told us we had accomplished more than any other team they had seen, which meant nothing to me given that I hadn't contributed a single useful idea. In fact, even the ideas I had contributed had been wrong, so they had actually slowed us down and impeded our progress. I left the operational interview with a newfound respect for the University of Illinois and a thoroughly shattered ego.

17

THE MOST IMPORTANT DAY OF MY LIFE began just like the previous one. Breakfast at the hotel, shuttle to the NSA facility, and an "A" for "applicant" badge. I filled out some paperwork and sat in the waiting room until my name was called.

"Hi, Tom, my name is Lauren. I'll be conducting your polygraph interview today."

Lauren was about thirty years old and quite attractive. In fact, several months later, I would end up discussing exactly how attractive I found her . . . during a polygraph examination with another attractive polygrapher.

But this was no time to reflect on Lauren's appearance. Seven yes-or-no responses were all that separated me from my dream job.

1. "Are you engaged in espionage against the US?"
2. "Are you secretly involved with foreign nationals?"
3. "Have you engaged in acts of terrorism against the US?"
4. "Have you ever mishandled US-classified information?"
5. "Have you ever committed a serious crime?"
6. "In the last seven years, have you used an illegal drug?"
7. "Have you withheld or lied about anything on your security forms?"

Lauren gave me the questions beforehand so there would be no surprises. She even told me the answers: "No. No. No. No. No. No. No." That's it. A parrot could pass this test.

She fastened a cuff around my right bicep to measure my pulse, two rubber wires around my chest to measure my breathing rate, and clips on my left index and ring finger to measure my perspiration. As I answered each question with a *no*, these devices would record my physiological responses. If I was telling the truth, supposedly, my responses would be normal. But if I was lying, my heart rate would increase, my breath would become shorter and quicker, and my fingers would grow wet with sweat.

"Some people look up ways to beat a polygraph on the internet before they come here," Lauren said. "If I think for a moment that you are employing one of these methods, I will end your test immediately."

I hadn't looked up anything on the internet about polygraphs, so her threat wasn't a concern for me. If I had researched polygraphs on the internet, I would have learned a number of fun facts. First off, your odds of successfully deceiving a polygraph increase if you actually believe you can beat it. As I recall, psychologists figured this out by giving a bunch of research subjects a twenty dollar bill and telling them they could keep it if they could convince the polygraph they didn't have it. One half of the subjects were told, "This machine is unbeatable." The other half were told, "You can beat this machine; it makes mistakes sometimes." In the end, it was a self-fulfilling prophecy: The second group had far better success than the first.

However, I didn't need to cheat on the test. I honestly hadn't done any of the seven things they were asking about. I just needed the machine to correctly identify me as "not guilty." Unfortunately, polygraphs have a terrible false positive rate—roughly one-third of the time when someone is being honest, it will register as a lie. This is why no employer in America can legally require an employee to take a polygraph—except the federal government, that is. It's too unreliable. But the federal government would rather reject one thousand qualified, trustworthy applicants than let one untrustworthy spy or terrorist through. After all, the cost of a mole is enormous, and there is no shortage of applicants for positions in the intelligence community. So, for their purposes, the polygraph works just fine.

We began the test with a handful of control questions to measure my resting pulse, respiration, and perspiration when I was telling the truth and when I was lying.

"Is your name Tom Rudelius?"

"Yes."

"Is today Monday?"

"No."

"Is today Tuesday?"

"No." An intentional lie, to establish my baseline response for deception.

"Is today Wednesday?"

"No."

And so on. Lauren asked if there was anything else she needed to know, anything I wanted to confess, before we started. I told her there wasn't. She sat down behind her desk, told me to look at the empty wall in front of me, and fired up the machine.

"Are you engaged in espionage against the United States?"

"No." I automatically scanned my memory to see if there was anything I might have forgotten to mention, anything I might have done that could possibly be construed as espionage. Nothing came to mind.

"Are you secretly involved with foreign nationals?"

"No." I immediately thought about my Filipino roommate. But no, I had already discussed him with the investigators. My mind flashed instead to the cute girl who sat next to me in my quantum mechanics class. Was she Russian? Or Polish? Certainly her family was originally from somewhere in that area, but was she an American citizen?

"Have you engaged in acts of terrorism against the US?"

"No." Again, nothing came to mind.

"Have you ever mishandled US classified information?"

"No."

"Have you ever committed a serious crime?"

"No." My mind raced to all the illegal things I had done in my life. The music I had downloaded. The time that woman had caught Steve and me throwing rocks at the Little League snack bar. The high school football games I had snuck into to avoid paying. None of these were serious crimes, right? But then again, who's to say what counts as serious?

"In the last seven years, have you used an illegal drug?"

"No." I thought of the handful of beers I'd had at frat parties, the champagne my parents had given me at New Year's, the allergy medication and albuterol inhaler I'd used when I was a kid.

"Have you withheld or lied about anything on your security forms?"

"No." This was the closest thing to a lie yet. In truth, I had been a little lazy in describing my places of residence because I'd gotten sick of the stupid form and had glossed over some unimportant details.

"The test is over."

I exhaled deeply. My life had flashed before my eyes, one petty crime and white lie at a time. Lauren excused herself from the room to go discuss my results with someone. I never found out exactly who that was.

When Lauren returned, she was angry.

"One of the questions is really bothering you," she spat. "And I think you know which one it is."

Again my mind began to race. *I shouldn't have been so lazy with that form.*

I explained my reasons for feeling guilty about the "withholding or lying" question, and she assured me that the NSA didn't care about that sort of minutia. They wanted to know if I had secretly lived in some place like North Korea and hadn't bothered to tell anyone.

"Well, you're right, that was the question that was bothering you," she said. She offered to reword the question on the test so that I wouldn't be hindered by passing over small details like this. I happily accepted the offer.

"Is there anything else you want to tell me about?"

I thought back to all that had gone through my mind during the test. There were a lot of little things I had done wrong, but none of it was serious enough to be of interest to the NSA.

"No, that's it."

She hesitated. It was clear she didn't believe me. She may have said there was only one question bothering me, but her look said, "You just failed every question on the test."

"OK, let's try again."

The machine fired up.

"Are you engaged in espionage against the United States?"

"No."

18

AFTER THREE REPEATED TESTS and three repeated failures, Lauren was downright furious. She picked up her chair from behind her desk, slammed it down in front of me, sat down, looked me in the eye, and shouted, "What aren't you telling me?!"

By this point, I was sobbing. I thought of all the questions I had answered in my life, all the tests I had aced without effort. And look at me now, unable to pass a test of seven yes-or-no questions, with all the answers given to me ahead of time. Worse yet, weeping like a little boy.

I realized that the test could not detect whether I was actually lying with my answers. Rather, it was going to fail me if I *felt guilty* about anything—at all. Anything that would cause my mind to start racing; any piece of my past I subconsciously wanted to keep hidden; anything about me that I was too ashamed to admit—all of it came flooding into my mind every time Lauren asked a question. I understood that the key to passing the test was to switch my brain to bypass—to turn everything off except for breathing and saying *no* seven times in a row. But the whole reason I had succeeded in school—the whole reason I even got the job offer from the NSA—was because my brain never shuts off. It's always whirring around at a million miles an hour, comparing every physics or math problem to the thousands I've seen before and looking for just the right one—the one that taught me the trick I would need to solve the problem at hand. And now, all of a sudden, one of my signature strengths had become my greatest weakness.

When Lauren asked if I'd ever committed a serious crime, I wasn't supposed to think back to all the little peccadilloes from my past and conclude that each one was too small to count. No, I was supposed to not think about anything at all. But that was impossible. As surely as the command, "Don't think about a pink elephant on roller skates," will conjure up that precise image, so too the command, "Don't think about anything at all," made me immediately think about anything and everything.

Thus, I had only one hope of passing the test: I had to make sure that my conscience was free of anything I might feel guilty about or embarrassed to share. Anything I held back was a possible dwelling place for my mind, a bomb that might detonate whenever Lauren asked the most closely related question.

Was it really worth it? I could just keep trying—maybe one time I would pass by sheer luck? That didn't seem very likely. The only other option was to drop out—tell Lauren I'd had enough and just wanted to go home. But this was my dream job. A few moments of painful honesty now would mean a lifetime of happiness later. There was only one way to solve this: I had to start talking.

And talk I did. I talked about everything, and I mean *everything*. From the little things—sneaking into high school football games, throwing a banana across the cafeteria in second grade, smuggling multiple oranges from the dining hall in violation of Cornell's strict "one piece of fruit or one dessert may be taken out of the dining hall per meal" policy—to the great embarrassments and failures of my life that I had wanted to keep buried forever. It was amazing how much dirt I turned up once I started digging.

Lauren—like all the polygraphers who would follow—decided that the fact I had never been on a date meant I must be gay. After the second or third time she questioned my sexual orientation, I got fed up.

"Look, I'm not gay! I mean, I don't have anything against gay people. I would tell you if I was, but I'm not!"

My denial was so strong that she actually wrote in my file, "Adamantly denies being gay." Later polygraphers would read this note and conclude that only a closeted gay person would deny being gay so vehemently.

Lauren kept probing, convinced I was hiding something enormous underneath my vast web of moral failures and minor crimes. Concerned I was holding back for fear of losing my shot at the security clearance, she started giving me examples of other people she had interviewed who had in fact gotten clearance despite terrible crimes in their past. One had committed murder. Many had hacked into various websites for fun. I have no idea if these stories were true or if they were merely fictions designed to lower my defenses, but it didn't matter to me. I told her everything that came to mind, but every time we reran the test, I became fixated on something new. The test would end, and I would have to confess again. Start. Stop. Fail. Confess. Repeat. The cycle went on and on.

Minutes became hours. The words of the woman who had given me my itinerary echoed in my head: "Four hours . . . because the test is rather stressful." She wasn't kidding. My test lasted right around four hours, and it was one of the most challenging and painful experiences of my life. I had been broken down and had my deepest secrets pulled out of me one by one. Yet, I walked out that day with my head held high.

I walked out knowing I was a Christian.

19

OVER THE PREVIOUS YEAR, I had become more convinced of God's existence. I had grown to appreciate Jesus as a great moral teacher and example. But I had still felt as if it wasn't all that important for me to become a Christian. I felt I was strong enough in my own mind to make it through life without religion.

Steve had tried several times to convince me that *everyone* is sinful and in need of forgiveness from God. I had trouble reconciling this claim with what I had observed to be true of myself and the world around me: As far as I was concerned, I was basically a good person, as were most of the people I knew. Sure, I wasn't perfect, but compared to my gifts and talents, my flaws felt like small dents that could be mended with a little bit of help and persistence.

Three hours into my NSA polygraph ordeal, I began to see myself in a new light. It was rapidly becoming clear that I didn't need "just a little bit of help." All the As on my transcript, all the hours of community service I had performed, all the accolades on my résumé, all the strong references from my friends—none of it was going to help me pass the polygraph. The polygraph is designed to find *any* problems with the applicant, not to weigh the good things they have done against the bad. And by that standard, I failed miserably.

It reminded me of the way Steve had described God's justice: Because he is perfectly just, he cannot allow even a single crime to go unpunished. If the standard for goodness were some serial killer, I'd come out smelling like a rose. If it were your average Joe, I'd probably still do fine. But if this is God's universe, and the standard is God's perfection, then I fall well short. It isn't just the murderers, the drug dealers, the corrupt politicians who need forgiveness: I, too, am in need of God's mercy.

The problem wasn't simply that I had done some bad things. The problem went much deeper than that. As I sat in the polygraph chair, I began to recognize the patterns of selfishness that underscored all my wrongdoings—both major and minor. I wasn't a good apple tree that happened to produce a few bad apples. There was a rottenness in me that went all the way to the core.

What I needed was not a little boost, but a clean slate, a new nature. As C. S. Lewis aptly noted, "Fallen man is not simply an imperfect creature who needs improvement: he is a rebel who must lay down his arms."[2]

The distinction had seemed insignificant to me before; but as I sat in that polygraph chair, it started to make sense. The things that Jesus was supposed to offer me—mercy, forgiveness, redemption, salvation, the Holy Spirit—these were all just religious buzzwords until I realized how much I needed them.

All of a sudden, the message that my brother had been trying to teach me made way more sense than it ever had before. In the words of pastor Timothy Keller, "We are more sinful and flawed in ourselves than we ever dared believe, yet at the very same time we are more loved and accepted in Jesus Christ than we ever dared hope."[3]

Christianity was no longer an abstract set of principles and doctrines; it was a story that made unparalleled sense of my own experience and a balm for the affliction that I hadn't realized I needed until that moment.

A year earlier, Steve had gone looking for hope of life beyond death, and he found it in the resurrection of Jesus. In that polygraph room, I suddenly found myself in search of forgiveness, and I found it in the crucifixion of Jesus: the moment when God's justice landed on Jesus so that his mercy could land on us; where God himself bore the weight of punishment so that guilty humanity could go free.

As I walked out of the NSA facility, I could almost taste that freedom. I had failed the polygraph miserably, yet I wasn't thinking about my poor performance, or the internship, or even the time I was about to spend with my dad in his DC apartment. I was thinking about the words Lauren had spoken as I had sobbed in her polygraph chair.

"Look at yourself, Tom. You are not the person you were when you walked in here. You walked in confident. You were sure of yourself. Now you can't even look me in the eye."

Lauren had no idea how right she was. I was not the person I'd been when I walked into that room. I had seen my brokenness play out before my eyes, one memory at a time. I had recognized my need for forgiveness. And in that moment, I turned to Jesus.

20

NOW THAT THE POLYGRAPH had revealed the gaps in my character, I felt compelled to fill those holes and to right my past wrongs as best I could.

I started thinking about people I had wronged in my life. The first ones I thought of were Dad, Mom, and Steve. As a teenager, I had often responded to my parents' love with bitterness, resentment, and ungratefulness. It was time to make amends.

Dad was living in an apartment just outside of DC at this point, having recently separated from Mom. He drove up to Linthicum and picked me up from my NSA interview. Our relationship had grown increasingly strained over the last few years. As college application time approached, he had pushed me hard to make sure I got into the best schools, and I responded by pushing back equally hard with sarcasm and disrespect. After we made it back to his apartment that day, I apologized to Dad for how I'd treated him as a teenager. He accepted my apology and responded with one of his own, saying he'd only ever wanted what was best for me and was sorry he'd gotten on my nerves in the process.

One time as a teenager, I had yelled at Mom that she was a terrible mom. She went to her room and closed the door. From outside in the hallway, I could tell she was crying—one of the very few times I had ever heard her cry. The afternoon of the polygraph, I called her and told her how much I loved her. I told her she was the best mom anyone could ever ask for. And she started crying and said she loved me too.

Steve and I will forever be each other's best friend. Yet there were lots of things we couldn't talk about when we were younger, lots of gaps that our relationship didn't cover. And one of those things was simply being able to say, "I love you."

"Hey Steve, I'm calling just to tell you that you're my best friend, you're the best twin brother anyone could ask for, and I love you."

"Umm . . . okay," Steve replied. He was clearly confused about why I was calling him and telling him this now. He probably figured something was seriously wrong. But the lack of an "I love you, too" in return didn't make the call less sweet. I had finally been able to say the words that I should have said to him a long time ago.

Some of the wrongs I had done could not be put right, but others could. The next day, I visited the mall near Dad's apartment, withdrew cash from the ATM, and walked into the Apple store.

"Hello, welcome to the Apple store. May I help you?"

"Yes, I have illegally downloaded some music in the past. I now regret this. Can you please take this money from me?" I held out the bills. The girl was completely bewildered—this was clearly not something they had covered in her training.

"Um, no, we don't just take money. But if you want to purchase some music *legally* in the future, you can buy one of our gift cards."

"Hmm, no thanks," I said. I walked out of the store disappointed that I couldn't just give them money without getting something in return.

But wait a minute! They could make me buy a gift card, but they couldn't make me use it. I did a 180 and walked back into the store. The girl was talking to her coworker. Evidently, she was talking about the crazy customer who had just tried to hand her a wad of cash, because when she saw me coming, she exclaimed to her coworker, "It's him!"

"Hi. Actually, I will buy that gift card from you."

I took the gift card home, cut it into pieces, and threw it in the trash. Mission accomplished.

Next, I visited my high school and stopped by the athletic department.

"Excuse me, I graduated from here in 2008. When I was a student, I snuck into a number of football games without paying. I now regret this. Can you please take this money from me?"

I held out forty dollars.

"Well, that's very nice of you. But you don't have to—"

"No please, I insist. Consider it a donation to the TJ booster club if you want—just please take the money."

She smiled. "Okay. You're a very honest young man."

The irony was not lost on me. The day before, I had failed a polygraph. Now I was being praised for my honesty.

In a column by my favorite sportswriter, Rick Reilly, I had read about the charity United to Beat Malaria, which provides insecticide-treated bed nets to fight malaria in Africa.[4] I visited the website and made a donation.

It wasn't until later that I noticed a passage in the Gospel of Luke about the encounter between Jesus and Zacchaeus, a corrupt tax collector. Zacchaeus climbs a sycamore tree just to get a glimpse of Jesus, who is passing by, and Jesus responds by asking Zacchaeus for a place to stay. Zacchaeus is so honored by this request that he says, "I will give half my wealth to the poor, Lord, and if I have cheated people on their taxes, I will give them back four times as much!"[5] To which Jesus responds, "Salvation has come to this home today, for this man has shown himself to be a true son of Abraham."[6]

The similarities between Zacchaeus's response to Jesus and mine were obvious—we both felt compelled to repay our debts, and to give to the less fortunate. Perhaps I was reading too much into these similarities, or perhaps it was a sign of my own insecurity in my newfound belief, but as I read Zacchaeus's story, I felt reassured that what I had just experienced was indeed a legitimate experience of God.

Along with my conviction to be a better person came a mysterious conviction that Christianity really was true. I felt a boost of confidence like nothing I'd felt before—not in my own abilities or moral fiber, but simply in the knowledge that everything was going to be okay because God was in control.

21

I WAS AFRAID TO TELL ANYONE about the failed polygraph for fear of accidentally saying something the NSA didn't want me to say and having to confess that at my next polygraph; but I wasn't afraid to tell Steve about my new faith. At 1:22 a.m. on January 15, I sent him an enigmatic, overly dramatic message:

> I consider myself a Christian now, not because I find it all scientifically, objectively reliable, but rather because I have positively changed so much over the last three days that I feel like only what you would call the "Holy Spirit" could be responsible for it. I am overflowing with confidence and a drive to

be a better person—for example, I just donated
$100 to nothingbutnets.net today. I guess I finally
understand John 6:44: "No one can come to me
unless the Father who sent me draws them, and I
will raise them up at the last day."[7]

Today, I cringe a little when I read that message—in part
because of the melodrama, but also because it runs counter to the narrative a religious scientist is supposed to tell.
Part of me wishes I could say that my decision to embrace
Christianity was a purely scientific one, based solely on intellectual grounds, which came after a thorough investigation of
the arguments for and against. Yet, although my conversion
experience came after a long and extensive investigation of
the arguments, the truth is that those arguments were ultimately insufficient at the time I heard them: My experience
of God played a crucial role as well.

On the other hand, it would be a mistake to say that I
decided to embrace Christianity in spite of my intellectual
beliefs, as if my decision were a leap from science into irrationality. Already, even before the polygraph, I had begun
to think of myself as a Christian. Given the choice, I would
have taken the side of theism over atheism in an academic
debate. The lesson from my conversion isn't that the arguments for Christianity are stronger than the arguments for
atheism, or vice versa. The lesson is that the arguments for
either side don't exist in a vacuum. They are always filtered
through the lens of the observer, even if that observer is a
supposedly objective scientist.

The Christian understanding of the world and human-kind as fundamentally broken hadn't resonated with my own life experience the first time Steve explained it to me. But now, on the other side of the polygraph, it did. As my experience changed, so did my view of the world, and this new view fit better with Christianity than it did with the worldview I left behind. The arguments for and against Christianity didn't change, but my evaluation of them did.

So, it wasn't that all my questions were suddenly answered, or that all the lingering doubts I had about Christianity went away. But over the previous few months, I had found myself drawing closer and closer to faith. My experience on January 12 wasn't foolproof evidence for Christianity, but it was enough to break the tie, to push me over the edge and compel me to action. I took the leap of faith, not from science into irrationality, not from uncertainty into certainty, but from uncertainty into what seemed like less uncertainty.

In the eight months since Steve first told me about his religious awakening, I had grown significantly in my knowledge of religion. At the same time, I had grown quite a bit in my knowledge of physics, as well, and it's possible that my deeper understanding of science made me more willing to live with some of the profound mysteries of faith.

Science, I was beginning to realize, had plenty of its own limitations, yet that didn't make science less remarkable. The inability of physicists to explain the big bang didn't give me reason to reject the rest of cosmology. The inability of biologists to explain the origins of life didn't give me reason to reject the rest of the evolutionary process. The remaining questions I had about faith seemed similar in nature: I was still puzzled by evil and suffering, God's sovereignty vs. human responsibility, and so much more, but now I felt that my inability to explain these great mysteries was no longer sufficient to justify an outright rejection of faith.

Steve, meanwhile, had spent those eight months trying to persuade me to embrace Christianity, and it seemed to him that I just wasn't very interested in it. So, when he woke up on January 15 to the announcement that I was now calling myself a Christian, he was shocked and thrilled. He immediately found Matt and showed him the message, weeping tears of joy as Matt read. Apparently, he didn't even try to hide his tears, as their floormate Mike could hear him crying from out in the hallway. He figured something terrible must have happened—perhaps a close relative had passed away? (Notably, Mike himself became a Christian six days later.)

Steve called me later that day. I was now at the Baltimore airport, waiting to fly to Minnesota for the rest of winter break.

"So . . . I got your message," he began.

"Umm, yep," I said.

"So would you say you're ready to put your faith in Christ?"

The question momentarily startled me. I felt that God had metaphorically reached out and touched me in that polygraph room and had given me a huge boost of confidence and peace. I liked that part of Christianity. But now I was being asked for something in return—to commit to following Jesus through hard times as well as good times; to lay my own desires aside and live life as a Christian.

I very seriously considered saying, "No, I'm still not ready to go that far." The words were on the tip of my tongue. But I just couldn't say them. God had just given me the revelation I had longed for: If I couldn't say *yes* to him now, when could I?

"Yes."

This was the decision that would henceforth define my life.

Sitting at the Baltimore airport, I prayed with Steve over the phone. He reminded me that the words I spoke weren't magic—it's not like saying the exact right combination of words is what will forgive our sins. But we can actually talk to God through prayer, and we can actually ask God to fill us up with his Holy Spirit to help us follow him and love like he does. So I prayed, thanking God for what he was doing in me and committing my life to him.

It was the end of a difficult journey to faith—but the beginning of a far more difficult path.

22

YOU MIGHT BE WONDERING whatever became of my time with the NSA. This was a journey on its own.

After I failed the first day of polygraph testing, they invited me back for a second one. I promptly failed it as well.

They invited me back for a third try. I failed that one as well. However, I did manage to pass the first four questions—they were now convinced I wasn't a terrorist—so I no longer had to answer those four, but I was still having trouble with the character-related ones—especially the one on "serious crimes." Whenever they asked that question, I could not keep my brain from dwelling on all the little, dishonest things I had done.

A couple of times, my polygraphers seemed worried that I might just be a pathological liar; so they added ten more questions into the next run to test their suspicions. I seemed to do just as poorly on those questions as I did on the others, but I don't think they gleaned any useful information from that. Evidently, they still couldn't tell if I was a liar or just a kid who was really bad at polygraphs.

Believe it or not, they invited me back for a fourth session. This goes to show how unreliable the test is. Even after three failures, each one involving a handful of tests, they still weren't convinced I was lying. The most NSA polygraphs I ever heard of someone failing was eight, and that person was finally awarded a security clearance after the eighth attempt. But I was only applying for a summer internship, which meant I had a deadline. If I couldn't pass by the start of the program in late May, I was out of luck.

I had some very strange conversations with the polygraphers. Once, when the tester was out of the room, I started thinking about an episode of the show *Arrested Development*, which made me start laughing. When she returned, she was as furious as I had ever seen—which is saying something.

"I want to know what's so funny that every time I leave the room, you start laughing? Is this a joke to you?"

I sheepishly told her I was laughing about a television episode I liked. That same woman also yelled at me for throwing my pen up in the air and catching it—a long-held nervous habit of mine.

Another polygrapher, after a long discussion and subsequent angry rant, grabbed a small piece of paper, drew a dot in the center, and stuck it on the wall in front of me with a pushpin.

"What's that?" he demanded.

Is this a trick question?

"It's a dot," I said.

"Yes, it's a dot," he fumed. "Stare at it." And he fired up the polygraph for another run.

They were convinced I was hiding something major about my sex life. No, once again, I'm not gay. No, I've never committed bestiality. No, I've never watched child pornography. No, I've never sold pornography. In fact, I've never even bought it.

"Oh, so you illegally downloaded it?" The polygrapher reached for a pen and pad of paper, suddenly intrigued.

"No, I'm talking about really soft-core stuff. Stuff that's freely available." I named some of the websites I had visited.

"Oh, why would you think that even counts as pornography?"

We spent a while debating the definition of pornography. I understood why US Supreme Court Justice Potter Stewart finally threw in the towel on this question and said, "But I know it when I see it."[8]

This description of the polygraph process might sound amusing, but I guarantee you it was anything but. It was torturous having to confess for hours at a time to a stranger who had no concern for my feelings and would scream at me when I started crying.

The polygraphs took a toll on me that extended far beyond the barbed wire fence of that NSA facility. The worst parts of me bubbled to the surface during each session, and I would continue to think about them for weeks after. My fragile ego couldn't handle it, my self-esteem plummeted, and I started thinking some dark thoughts. One time, I was walking down the streets of Ithaca, and a car stopped to let me cross at the crosswalk. I realized that if I didn't exist, that car wouldn't have had to stop for me; the driver could have gotten where he was going slightly faster. And I wondered if my measly existence was worth more than those few seconds of his time. Maybe it would have been better if I hadn't been born?

Sleep deprivation took a toll as well. I never pulled an all-nighter in college, and I have strong philosophical views against the practice. But the nights before the polygraphs came close to being all-nighters. It wasn't unusual for me to have trouble sleeping, as a college student. Add to this the anticipation of the following day's test, and I was happy if I fell asleep at all. The 8:00 a.m. start time for the polygraph didn't help matters either. I had to set my alarm for no later than 6:00 a.m. in order to catch the shuttle to the NSA facility.

Having to go through this process all alone made it even harder. I was still too afraid to tell anyone about the polygraphs. All my parents knew was that I kept being called back for more interviews. The spacious hotel room where I stayed during my visits to Linthicum only reminded me how alone I really was in the midst of all this. What I would have given to have Steve there to encourage me and wish me luck each time before the following day's test.

In this isolation, I started to pray a lot more. I trusted that God had a plan for me in this and that he was going to stick with me until the end. And that mindset was crucial, especially after learning I had failed my fourth polygraph. By now, we were less than a month away from the deadline, and up to this point my polygraphs had taken roughly a month to schedule. Was it too late to try one last time?

23

MEANWHILE, I WAS LEARNING more about my newfound Christian faith. On January 18, 2010, not long after I had prayed with Steve on the phone in the Baltimore airport, he mailed me a letter:

Tom,

What an incredible joy it was for me to hear of your accepting Christ into your life! It truly is the best decision you could ever make in this world, and I am so glad and so happy for you!

I'm not exactly sure what kind of spiritual journey you went through to make it to this point, but I know you had very earnestly been seeking the truth about God, and I believe that "he who seeks finds" (Matthew 7:8), and "if you seek him, he will be found" (1 Chronicles 28:9). Therefore I want to thank God for building up in you a desire to seek, for guiding you in your search for truth, and for opening your heart and mind to believe and receive the love that Jesus died in order to give you. It is so amazing to think of what God has done in your life in a relatively short period of time, and I am so excited to see what He will continue to do.

Speaking of continuing in your walk with God, I have decided to send you a few things that might prove valuable. I believe you've seen both of the booklets before: The first, known as the KGP [Knowing God Personally], is all about entering and growing in a relationship with God, and the second is about leading a Spirit-filled life. They both do a very good job of communicating the Christian faith, and the KGP especially is an extremely useful tool for summarizing the gospel message in four quick points. I have also sent you a Bible verse to remind you to keep your life centered on Christ.

A few things I want to encourage you about:

At this point you've probably read more of the Bible than most Christians have—but don't stop now! Second Timothy 3:16-17 [NIV] says, "All Scripture is God-breathed and is useful for teaching, rebuking, correcting and training in righteousness, so that the servant of God may be thoroughly equipped for every good work." Continue to strive to know Christ deeply through His Word.

As I said on the phone, I think you need to get involved in some sort of Christian group on campus. . . . One of the most important aspects of the Christian faith is that it's relational—God works through people to reach other people. As children of God, we have been called both to meet with other Christians (Hebrews 10:25) and to reach out to nonbelievers with the gospel message (Matthew 28:18-20). I don't see how these things are possible without some sort of Christian group in which to grow. I know it must be hard trying to get involved somewhere when you probably don't know many Christians around, but use this challenge as a way to grow in your faith, by trusting that God will lead you to a group that you can be involved in. If you need help, my mentor Tim says he knows a few people who graduated from Cornell—I'm sure they'd know people there who could help get you plugged in somewhere if you want.

I think you should get baptized. Not because it will save you, for it is through grace you have been saved, not by works; but rather because it is a way for you to tell people of the change that has happened in your life. It may be sort of awkward to say to people, "I'm a Christian now." It could very well be a whole lot easier to get baptized, as a way of publicly displaying your faith and allowing others to witness the works God has done in you. If you want to hold off on this for a while, I understand. Obviously the first step is for you to find people at Cornell that you can connect to. But one day you might find that God is calling you to get baptized, and I would just encourage you to keep that possibility open in case He does.

Well that's good for now. Again, I'm so happy to see you take the step to place your faith in Jesus—you will find, as I am constantly learning, that God will transform you in so many ways that you never would have dreamed possible. I have always admired so many things about you—you are the only person I know who could give me an answer every time I was stuck on a math problem—you are the only person I know who could give 100 percent effort to a last-place baseball team—and you are the only person I know who could so effectively memorize *Seinfeld* quotes and use them to produce hilarity in real life. God indeed created you personally and has a wonderful plan for your life. Now that you're part of His Kingdom, I can't wait to see how He'll use you to carry out that plan.

Don't hesitate to call me if you have any questions. Remember always that I love you and I am here for you, and that I am praying for you daily.

In Christ's Love,

Steve

Along with this letter, Steve included some verses from the Gospel of Luke:

One day Jesus told a story in the form of a parable to a large crowd that had gathered from many towns to hear him: "A farmer went out to plant his seed. As he scattered it across his field, some seed fell on a footpath, where it was stepped on, and the birds ate it. Other seed fell among rocks. It began to grow, but the plant soon wilted and died for lack of moisture. Other seed fell among thorns that grew up with it and choked out the tender plants. Still other seed fell on fertile soil. This seed grew and produced a crop that was a hundred times as much as had been planted!" When he had said this, he called out, "Anyone with ears to hear should listen and understand."

His disciples asked him what this parable meant. He replied, "You are permitted to understand the secrets of the Kingdom of God. But I use parables to teach the others so that the Scriptures might be fulfilled:

'When they look, they won't really see.
When they hear, they won't understand.'

"This is the meaning of the parable: The seed is God's word. The seeds that fell on the footpath represent those who hear the message, only to have the devil come and take it away from their hearts and prevent them from believing and being saved. The seeds on the rocky soil represent those who hear the message and receive it with joy. But since they don't have deep roots, they believe for a while, then they fall away when they face temptation. The seeds that fell among the thorns represent those who hear the message, but all too quickly the message is crowded out by the cares and riches and pleasures of this life. And so they never grow into maturity. And the seeds that fell on the good soil represent honest, good-hearted people who hear God's word, cling to it, and patiently produce a huge harvest."

I was already planning to read the rest of the Bible— namely, the Old Testament—so Steve's first suggestion wasn't anything new. I agreed that baptism would be a good idea at some point, but it didn't seem urgent. It was Steve's second suggestion that really challenged me: If I was going to continue to grow as a Christian—if I was going to be that "seed that fell on good soil"—I had to get involved with a Christian community.

24

BY NOW I HAD MET A FEW NICE, relatively normal Christians my age, but there was no guarantee that the Christians I would meet on campus at Cornell would fall into this category. I still attached a stigma to kids who had been raised religious, and I was also afraid I wouldn't fit in. Plus, I wasn't sure I wanted the hassle of going to church; I had other things I wanted to do on Sunday mornings.

So, for a while, my attitude toward Christianity was like that of Chris Tucker's character, LAPD detective James Carter, in the film *Rush Hour*: "I don't want no partner, I don't need no partner and I ain't never gonna have no partner."[9] I was going to be a Christian without anyone's help.

This perspective is pretty common. Many people have

had experiences with church ranging from disappointing to terrible, but they don't want to give up on faith entirely. Years later, when I was a postdoc at Princeton, I mentored a college student whose priest had been charged with molesting children. After that, the student's family got together and decided they were going to follow Jesus on their own—no more church. I didn't blame them.

But let me be honest: I didn't have a good reason for avoiding church. I hadn't been attacked, abused, or otherwise mistreated by Christians or the church. I was just afraid of the unknown.

The way I finally got involved with Christian community was through intramural basketball. A couple of my friends played once a week at the rec center with a guy named Nic. Nic was in his late twenties and happened to be on staff with a Christian group on campus. When intramural basketball season came around, our mutual friends started a team and asked both of us to be on it.

Nic and I didn't talk about Christianity at all during the season. We were focused on trying to win games, and I hadn't summoned the courage yet to talk about deeper things. Nic was very good at basketball—definitely the best player on our team—but the rest of us were mediocre at best, and we got knocked out early in the playoffs.

As Nic later told me, he was sitting in his car before our final game, wondering why he was playing on this team. He had a wife and a couple of toddlers at home and was stretched thin with his job in campus ministry already. As he sat there, he prayed that God would give him an opportunity to talk with just one guy, to explain his faith and why it meant so much to him.

Evidently, I was the answer to his prayer. After the game, I finally worked up the courage to talk to him about my new faith.

"Umm, hey Nic, you're on staff with a Christian group here on campus, right?"

"Yeah, bro, that's right," he said. (Nic says "bro" a lot.)

"So, I just became a Christian a couple of months ago, and I've been meaning to get more involved here on campus."

Nic invited me to join the small group Bible study he led on Monday nights, and I attended for the rest of the semester.

Nic's group was small—about five of us in total, but only two or three would attend in a given week. One of the guys would often take over the conversation and send it in bizarre directions. But I'd had no idea what to expect, so I was pleasantly surprised that it wasn't even weirder.

I was also pleasantly surprised to find that my knowledge of the Bible stacked up pretty well with the other people in the group. Even though I hadn't grown up attending church and Sunday school like they had, my reading of the New Testament, my conversations with Steve, and the books about Christianity he had given me were more than adequate as a primer. I was able to contribute to group discussions and occasionally even impress Nic and the other students in the group by how much I knew about the Bible. It was a far cry from how I'd felt at Steve's baptism, where I had felt like a total outsider around his group of Christian friends.

At the same time, I was impressed by how Nic and some of the more seasoned Christians in the group were able to apply their faith to their lives, even in the most difficult situations. One student showed up to group one day and told us that his grandfather had just passed away. At the time, I was feeling overwhelmed by the grind of the Cornell semester and the weight of the polygraph process, and the thought of losing a loved one was almost unbearable. But this student seemed genuinely at peace with his grandfather's passing, confident that his grandfather was now with God.

The small group also gave me a chance to know Nic better, and I quickly realized he was one of the most faithful and genuine people I'd ever met. At some point he started meeting with me one-on-one to teach me about Christianity and help me grow spiritually. I probably could have continued to believe intellectually in Christianity without Nic's help, but when it came to the practical question of how to live faithfully as a follower of Jesus, I grew far more with Nic as my role model and coach than I ever could have done on my own. Nic was simultaneously encouraging and affirming, yet he was also willing to tell me that I was wrong. (And as a young, hardheaded know-it-all, I was wrong a lot.) But he always did it with love.

More than anyone else I knew, Nic could tell that my ongoing interview process with the NSA was taking a toll on my mental health. This was in part a testament to the amount of time he spent with me that semester, and partly due to his ability as a campus minister to spot when a student wasn't being completely forthright with him. In his typical straightforward fashion, Nic asked me one day what was bothering me.

"I'm not supposed to tell you very much," I said, "because I'm trying to get a top secret security clearance for an NSA internship, and I'm not supposed to talk about the application process. But suffice it to say that it's been really tough, a lot tougher than I expected, and I might have already blown my chance at it."

Nic tried to push further, but I wouldn't tell him anything more. So he simply reminded me about God's love for me and prayed with me.

I left our conversation thankful that I'd spoken with him, but also paranoid that I might already have broken some NSA confidentiality clause by mentioning the security clearance process at all. On the other hand, I wasn't even sure that the NSA would invite me back for another interview, so part of me regretted not just telling Nic everything and unburdening myself from the weight of secrecy.

The next day, I was given a glimmer of hope. The NSA quickly scheduled a fifth polygraph to give me one last shot at securing the clearance. But this was clearly my last chance—one last installment of Tom vs. the polygraph, next basket wins, winner take all. Would I seize the moment, or let it all slip away?

25

THE FINAL ROUND OF POLYGRAPHS started with an operational interview—essentially a polygraph without the machine. Four hours of confession. They wanted me to get *everything* out of my system so I could go in and pass the polygraph.

It didn't work. When I came back the next day, they hooked me up to the machine and asked me the same, all-too-familiar questions.

"Have you ever committed a serious crime?"

"In the last seven years, have you used an illegal drug?"

"Have you withheld or lied about anything on your security forms?"

Once again, I failed. And once again, it was the serious crimes question that got me.

I started confessing again, but now I was the one growing impatient. I just wanted to take the test again and again until I got it right. This was my last chance, and I had only four hours to get my act together and earn that stupid security clearance.

After the second failed attempt, I made the mistake of mentioning that I had told "a bunch of lies" throughout my life. This startled my polygrapher.

"A bunch of lies?" the polygraph operator asked skeptically.

I tried to explain that everyone tells lies—far more frequently than they would like to believe. This was something I had realized in the past few months—you have no idea how often and how naturally you tell little lies to control others' perceptions of you until you have to take a series of polygraphs, and you subsequently learn to analyze every little statement you make to ensure its veracity. According to a frequently played commercial at the time for the TV show *Lie to Me*, the average person tells three lies for every ten minutes of conversation.

The polygrapher didn't buy my "everybody does it" excuse. She added ten extra questions to assess my character. Three questions away from a security clearance had now become thirteen questions away.

But something different happened this time when she added the questions: My fixation switched away from serious crimes to focus on one or two of the new questions. Now I was more concerned with little moral failures and social miscues. I started having trouble with those questions and stopped being so nervous about the serious crimes question.

The good news was that the NSA evidently doesn't much care about small moral failures—they acknowledge that no one is perfect. All they really want to know is, "Can we trust this person with secrets of national security?" So when my results came out, the polygrapher had surprisingly good news.

"You did much better on the serious crimes question that time. You struggled a bit on a couple of the others, but security may or may not have an issue with those. Overall, you should be proud of the work you did today—you did far better today than you have in the past. Good job."

All right! It was the first time a polygrapher had ever told me, "Good job." And it meant I was finally done with polygraphs. All in all, including my most recent operational interview confession session, I had spent about twenty-one hours in a room talking to an NSA agent and a camera. I had lost a ton of sleep over the course of several months and been pushed to the brink psychologically. The only question left was whether it was all enough to merit a security clearance.

I didn't have to wait long to find out. The next day, I was sitting in the computer science building, grading final exams for the course I was TA-ing, when I got a call from a restricted number with a Maryland area code. I excused myself from the room.

"Hello?"

"Hi Tom, this is Lisa, calling from the National Security Agency. How are you today?"

"Good. How are you?" I held my breath.

"I'm good, thanks. I have some good news for you. I just got word from security. Your clearance has come through . . ."

I didn't hear the rest of her words. I was already on cloud nine. All that hard work, all the grief I had taken from those polygraphers, it had all paid off! I had done it!

I prayed quickly before I went back to grading exams, thanking God for my success. As soon as I finished grading, I called Steve, Dad, and Mom, and sent thank-you emails to the professors who had written my letters of recommendation for the internship, letting them know I had gotten the job at last.

Four days later, it all came crashing down.

"Hi, Tom. This is Lisa again, calling from the National Security Agency. We've just gotten word from security. Your clearance has been rescinded. They can't tell us why. This has never happened before . . ."

Once again, I didn't hear the rest of her words. Nor, to be honest, could I have told you precisely what "rescinded" meant before that conversation. But I didn't need to look it up. It was clear from the context. I fell backward in a heap onto the carpeted floor of my apartment and lay there for a long time, struggling to understand what had just happened and what it meant for my life.

I finally got up. I was too ashamed to call my parents. I knew that if I heard their voices, I would start crying. Instead, I just sent them an email:

> I just got a call from the NSA, saying my security clearance has been revoked. I don't know why. I'm not sure what I'll be doing this summer—I may stay in Ithaca. I'll let you know.

I had to tell my roommates, my friends, my family, and everyone else who knew about the job that my celebration the previous Thursday had been premature. It was one of the hardest, most embarrassing things I've ever had to do. To this day, I still don't know what caused them to change their minds, or why anyone from the security department was even looking at my file after they cleared me. I was told that if I inquired, I could see the polygrapher's notes on me, but I didn't want to. I could already narrow the reason for the revocation down to a few possibilities—most likely the "bunch of lies" comment I had made. But the last thing I needed was to dwell even longer on my mistakes and regrets. I hadn't gotten the job, and that was that. I would never be a secret agent. It was time to move on.

I decided to focus on my faith instead.

Part III

Faith and Doubt

26

AS ANYONE WHO HAS EVER BECOME a Christian knows, when I came to faith, I didn't just live happily ever after. Over the past dozen years or so, I've experienced seasons of intense doubt and anxiety. At times I've asked myself, "Do I actually believe all this stuff?" My theological views have shifted significantly in certain areas.

I'll admit I was tempted, in writing this story, to make it seem as if all of those doubts hit me on my initial journey to faith, and that I resolved them all way back when. But chances are you already think I'm a liar, given all the failed polygraphs, so just to prove you wrong, I'm going to tell it the way it really happened.

I wasn't sure how my family would respond to my new-found faith. It turns out that it varied quite a bit from person to person.

When Steve became a Christian, our grandma thought he had joined a cult. When Mom got home from his baptism, Grandma asked her, "Are you okay with everything that went on down there?" (As if Steve might have been brainwashed, and his family needed to save him.)

Mom, on the other hand, was remarkably positive about our newfound faith. After my first polygraph in January 2010, I went to her home in Minnesota for the rest of winter break. At some point, Mom found out about my recent charitable donations and asked me what had led me to give financially like that.

"Well," I said, "I guess I decided that I'm a Christian."

"Oh, wonderful!" she exclaimed. And it was a genuine exclamation. Mom isn't typically the most emotional person, and she's a terrible liar, so when she exclaims something like that, you know she really means it.

Around this time, Mom also started going to church, which was something she hadn't done before during my life-time. I'm not sure exactly what caused her to re-explore faith, but I think my spiritual awakening—along with Steve's—played an important role.

Dad was, and still is, more hesitant about my faith. I don't remember when I first told him I had become a Christian, and I don't recall his reaction as especially positive or nega-tive. But early on in my faith journey, he and I had a num-ber of conversations (read, *arguments*) about God. And in my newfound zeal, I wasn't especially gracious during those conversations.

It seemed no matter what I said—any statement I made in favor of Christianity, God, or the Bible—Dad would bring it back to his own baseline view: The authors of the Bible lived in a time when people were at the mercy of the world around them for food, safety, etc. So they developed religious beliefs as a way to cope with uncertainty and to feel as if they had some sort of control over their uncontrollable circumstances.

Nowadays, this sort of Freudian reasoning doesn't faze me. For all I know, Dad's right—I don't know how our ancestors developed their belief in God, and I don't find the question especially relevant. How people in the past came to believe in God is far less important to me than the question of whether it's all true, and whether my own reasons for belief are sound.

But back then, I felt like I couldn't concede a single point. To do so would be to admit that my reasoning was flawed, my faith was invalid, and ultimately I was just a kid who didn't belong at the adult table, theologically speaking. So I tried to respond to everything Dad said, making counter-arguments that were not very compelling—to him *or* to me.

Despite our arguments, Dad was remarkably supportive of Steve's and my faith. I have several friends, raised in churchgoing, Bible-believing Christian households, whose parents were furious when they decided to pursue a career in ministry, believing they were wasting their talents and education and should pursue a more financially rewarding career instead.

When Steve decided he wanted to work full-time with an on-campus ministry, Dad was actually quite positive about it. Whatever his views on our faith might have been, he wanted us to live happy and meaningful lives more than he wanted us to fit into a particular lifestyle he might have chosen for us. This was also true for our extended family, more broadly, but not without some bumps in the road. Both Steve and I—enthusiastic about our emerging faith, but not yet wise enough to know when to keep our mouths shut—had a fair number of theological arguments with grandparents, aunts, and uncles in our first few years as Christians. We found that most of our family also considered themselves to be Christians, but their theological views were often very different from ours. Some of my relatives, like many of my peers, held a relativistic view of religious truth: "What's true for me may not be true for you." "Maybe miracles exist in your reality, but they don't exist in mine."

As a scientist, relativism has never appealed to me. The goal of science is to search for truth about the world. We may never arrive at 100 percent certainty, but a lack of certainty about reality is not the same as the absence of reality. We all acknowledge that there are laws of physics, even if we don't always know what they are. We acknowledge events in history, even if reasonable people might disagree about what exactly transpired. To me, religious truths are either in the same category as scientific truths and historical truths—objective realities about our world—or they're not worth my time.

This latter perspective on religious truth is what I found so appealing about Christianity in the first place. The truth of Christianity, as Steve explained it to me, hinges entirely on whether or not Jesus rose from the dead. It's as much a question of history as it is a question of religion. On the other hand, the religion practiced by many people could survive even if Jesus had never existed. To some members of my family, practicing the moral code that Jesus espoused ("love your neighbor as yourself") and observing certain rituals (going to church on Sundays, celebrating Christmas and Easter, and so on) are the hallmarks of the faith.

Back then, I was irritated that my family would use the word *Christianity* to describe a belief system that was more or less indifferent to the question of whether or not Jesus was actually the Son of God, the Messiah. Why celebrate Easter if you don't believe Jesus actually rose from the dead?

Eventually I came to realize that these sorts of debates don't actually change anyone's mind. My relatives didn't care how convincing my arguments were: Their minds were already made up. But then again, wasn't I doing the same thing? Maybe I was right about the importance of the Resurrection, and my grandfather was also right about the importance of humility, love, and Christian charity. Maybe Jesus is both a savior *and* a teacher, both Lord of all *and* a great moral example?

Once I started looking for common ground, it wasn't hard to find. One day, my grandfather printed out an excerpt from David Brooks's book *The Road to Character* for all his grandchildren to read. Though my cousins seemed relatively uninterested, my grandfather and I bonded over our shared appreciation for Brooks's wise words about the importance of pursuing holiness over happiness.

When I was a baby, I was baptized in my grandparents' church, which featured ancient religious rituals and traditional Christian hymns but seldom used the name of Jesus. When I first came to faith, I started attending a more modern, Jesus-focused church that featured contemporary worship songs and was almost entirely devoid of traditional Christian rituals or sacraments. Today, I attend a church in my grandparents' denomination that is centered on Jesus, sings a mixture of old hymns and contemporary worship songs, and practices many of the ancient rituals that my grandparents' church did. Perhaps there was more to their faith than I realized as a child.

27

ONE OF THE FIRST DECISIONS I made as a Christian was to take a college course on the New Testament. Steve encouraged me, but he also sent me a message, cautioning that my professor might have an unfavorable stance on Christianity.

Sounds like a good class to me—learning about the New Testament is always a good thing. I should warn you, though—the vast majority of religion professors are practical atheists who teach from the perspective that religion is a worldly thing, and that we can't know any sort of truth about God. Therefore, if you do take this class, make sure to pay special attention to the facts of what you learn

and don't be led off track because of the opinions
of your professor. Just remember that there are very
solid arguments in favor of Christianity, and that
people much smarter than you or I, or probably the
professor of that course, hold firm to the logic of
Christian beliefs. If you find yourself doubting what
you've learned about Christianity, that is not a bad
thing—I am confident that your doubts will lead to
answers that point to Jesus as God—but if you are
confused at all in this class, please call me.

I was so excited by my newfound faith that I really didn't
think much of Steve's words. Besides, I was never one to turn
down a challenge.

I realized pretty quickly that Steve's prediction was
right. The textbook for my course was the famous *A Brief
Introduction to the New Testament* by Bart Ehrman—a former
fundamentalist Christian turned skeptic, who is well known
for his debates against Christian apologists and his popular
books arguing against the divinity of Jesus. The course was
taught by a former PhD student of Ehrman's, who clearly
shared his skepticism toward the New Testament.

The teachings from that course will be familiar to anyone
who has taken a similar course: None of the four Gospels was
actually written by the people to whom they are commonly
attributed (Matthew, Mark, Luke, and John); only seven
of the thirteen letters attributed to Paul can confidently be
ascribed to him; Jesus never claimed to be God; and miracles
are scientifically impossible.

Such courses are notorious for destroying the faith of young Christians, many of whom have been raised to believe that every word in the Bible is literally true, only to be taught by their college professors that much of it is inaccurate, inconsistent, and impossible. At least one of my Christian friends dropped the course.

My own faith was strangely unaffected by these arguments. I suppose I was still riding the emotional and spiritual high of my conversion, and my own personal experience with God was so strong that my faith never wavered. I certainly never thought about dropping the course. From the very beginning of my spiritual journey, my goal had been to find the truth. I didn't want to believe in something just because it gave me comfort and hope. I wanted to believe in it because it was true. And I knew that if my faith couldn't withstand the criticisms of my professor and the textbook, it wasn't a faith worth having in the first place.

That said, the course definitely shaped the way I approached my faith. Many Christians—including a young Bart Ehrman—base their faith on the perfect and unerring veracity of Scripture. This means that if there's even one mistake in the Bible, the whole thing is called into question.

I realized quickly in that class that the inerrancy of Scripture could not be the linchpin of my faith. I had neither the knowledge nor the desire to try to counter every point my professor made against the trustworthiness of the Bible. Perhaps 1 Timothy really wasn't written by Paul, and later Christians mistakenly believed that it was and put it in the biblical canon—would that invalidate the truth of Jesus and his resurrection? Would that invalidate my own spiritual experience of God?

Paul himself realized that what really matters is the resurrection of Jesus (1 Corinthians 15:14-23). If Jesus rose from the dead, then the world is a magical place: Miracles happen; death has been conquered; and joy, beauty, and love are not merely convenient names for chemical reactions in our brains, but rather integral parts of the world around us. If Jesus rose from the dead, the rest of the Bible is just an added bonus.

Though my professor had a lot to say against the Bible as a whole, I couldn't help but notice that she didn't have much to say against miracles and the resurrection of Jesus. One day, she put up a slide of "almost indisputable facts" about Jesus of Nazareth, according to one scholar. There certainly wasn't a lot on it—Jesus was baptized, Jesus was crucified, etc. But at the end of the list was a rather surprising one: Belief in Jesus' resurrection was based on experiences of the resurrected Jesus by his followers. One student in the class was so surprised that he asked the professor about it.

"Umm," she said. "I'm not sure about that one." It seemed almost as if she had never thought about it before.

People in similar classes at other universities have had similar experiences. I once met a Yale graduate who, several weeks into a college course criticizing the New Testament, approached the professor after class to ask what happened on the third day after Jesus' crucifixion—the first Easter Sunday. The professor took a step back and spread his arms for emphasis: "Something *cataclysmic*," he said.

At Harvard, many young Christians lost their faith while

taking the New Testament class taught by Helmut Koester. Yet Koester once wrote:

> All that is certain is Jesus' death on the cross, for which the Roman authorities bear full responsibility. We are on much firmer ground with respect to the appearances of the risen Jesus and their effect. . . . Paradoxically, our most immediate testimony comes from a man who had not known Jesus personally: Paul. However, that Jesus also appeared to others (Peter, Mary Magdalene, James) cannot very well be questioned.[1]

Across the Charles River, at Boston University, a close friend of mine received what he said was probably the worst grade of his life on a paper he wrote for a class taught by the renowned historian Paula Fredriksen—his arguments clearly hadn't met her rigorous standards for historical evidence. Yet, in her book *Jesus of Nazareth, King of the Jews*, Fredriksen acknowledges that "the disciples' conviction that they had seen the Risen Christ . . . is [part of] historical bedrock, facts known past doubting about the earliest community after Jesus' death."[2]

I was surprised by the way that Bart Ehrman, in a sidebar in his New Testament textbook, tries to evade the question of miracles altogether:

> Today miracles are understood as events at odds with how the natural world typically works, based

on scientific understandings developed since the Enlightenment.

This means that miracles are, necessarily, the most improbable of events.

Since historians can establish only what *probably* happened in the past, they cannot show that miracles happened, since this would involve a contradiction—that the most improbable event is the most probable.[3]

Ehrman's argument didn't strike me as very compelling. Sure, I agree that miracles are the most improbable of events *if atheism is true.* But if God exists, and if Jesus Christ really is the Messiah, then wouldn't I expect to see some miracles? If I assume that Christianity is false, then of course the miracles of Jesus seem unlikely. But isn't the whole question whether or not Christianity is true in the first place?

The more I thought about Ehrman's logic, the more I started to wonder why this argument appears in his textbook at all. In fact, beyond the sidebar, he dedicates five pages to the issue.[4] Imagine if you were reading a textbook on Greek mythology and all of a sudden the author called a time-out to make the case that the stories of the Greek gods and goddesses didn't actually happen. Wouldn't you be puzzled? Of course they didn't actually happen! Of course Hercules never defeated a bunch of man-eating birds with beaks made of bronze and feathers made of metal! Of course Hera didn't pluck out Argus's eyes and stick them on the tail of a peacock!

I came to realize the reason is pretty simple: The evidence for the miracles of Jesus is far stronger than the evidence for the miracles of Hercules and Hera. The latter can easily be denied on historical grounds. The former requires us to ask questions of metaphysics. If your metaphysics doesn't allow for miracles, then of course you must deny the miracles of Jesus, as Ehrman does. But if there's room for God in your fundamental understanding of reality, then the case for the miracles of Jesus is actually quite strong. To quote the late atheistic philosopher Antony Flew, "If we were in a position to suppose [there is a Power external to the world], then no doubt the case for the occurrence of these particular miracles, as well as for that of the supreme miracle of the Resurrection, would be open and shut."[5]

In the end, the New Testament course led me to question some of the more peripheral parts of my faith (Was Jesus *really* born in Bethlehem?), but it bolstered my faith in other areas—most notably, the Resurrection. Even more important, it forced me to read and internalize the Bible. As someone who missed out on all the Sunday school lessons my churchgoing friends had heard while growing up, I felt—and occasionally still feel—as if I had not been privy to some important information and rites of passage. (On the other hand, all my Christian friends who went to church when they were kids seem to love *Veggie Tales*—and I don't quite get it.)

After that class, I found I had gained the confidence of knowing that my faith had been tested—and had survived. A faith that crumbles under scrutiny is not a faith I'm interested in anyway.

28

MY FIRST COUPLE OF YEARS OF GRAD SCHOOL were rough. I had excelled as an undergraduate, and I expected to do the same at Harvard. It turns out that things weren't nearly as easy as I expected them to be.

As an undergraduate physics major, your job is basically to beat the other undergraduates around you. All the problems you work on have answers, they've been solved by thousands of other people before you, and if you can't figure them out, you can always go talk to the professor or the TA, and they'll help you.

As a grad student in theoretical physics, however, your job is to do *original* research, which means you're trying to solve problems that no one has ever solved before—including some of the smartest people in the world, who have decades of expertise and research experience under their belts. I thought I knew a lot about physics when I started grad school. I quickly realized that I basically knew nothing.

My adviser, Cumrun Vafa, was known for taking a laissez-faire approach to his younger grad students, allowing us to flounder for a while as we learned how to swim. "Eventually," his older grad students told me, "things get good."

Cumrun is a dominant force in the field of string theory, he works hard to ensure the success of his students, and his former grad students have exceptional track records. So, once you've earned your standing as his student, you can be assured he'll go to bat for you, and the rest of the field will respect you all the more for having studied under him. But first you have to earn your standing as his grad student by completing a project—and I struggled with that, big-time.

The first four projects Cumrun gave me to work on didn't go anywhere—either because I was too slow and got scooped by some Japanese physicists who beat us to the result, or because I wasn't smart enough to know how to even get started.

I began to wonder if I had made a mistake by going to grad school in physics. Perhaps I wasn't good enough to make it in such a cutthroat field? Or was I wasting my time? I'd been captivated by theology and a desire to preach. I started to wonder whether I should just leave physics and pursue a career in ministry.

But I also knew that I couldn't give up on physics so easily. As I neared the end of my time in college and considered various career paths, I concluded that grad school in physics gave me an opportunity to use my unique gifts and talents to talk about my newfound faith with others. For whatever reason, our culture seems to care a lot about what scientists think about God. I dreamed of giving talks about science and faith, debating famous atheists, and writing a book about my journey to faith. My motives were undoubtedly mixed. I cared far too much about my own reputation and fame (and to be honest, I still do). But a part of me also understood that my career was more than just a way to make money or a way to achieve personal happiness and satisfaction: My career was a way to serve God and to help others experience some of the hope and joy I had found in coming to faith.

The feeling that I was working for something larger than myself helped sustain me in those tough early days of grad school. I spent many late nights trying to understand esoteric physics papers without any immediate gratification, and as months passed without any major breakthroughs in my research, I needed to somehow find the motivation to keep going. In those moments, I often glanced at a plaque on my desk that Nic had given me as a college graduation present. It contained a quote from C. T. Studd: "Only one life, 'twill soon be past. Only what's done for Christ will last." My faith gave me the strength to keep pushing.

In turn, I'd like to think that my career helped my faith. When I was in college, a fellow student once predicted that grad school in physics would surely destroy my faith. But he didn't realize that my faith was one of my main motivations for going to grad school in the first place. I probably would have pursued a different career path if I hadn't become a Christian. So every day in the office was like a tiny reminder of my purpose in life, and a reason to persist in my pursuit of God.

After Steve graduated, he decided to work full-time in campus ministry at his alma mater. Sometimes, people would point out to me the irony of one twin working in the field of science while the other worked in the field of religion. But I categorically rejected the idea that my work was opposed to my faith, drawing inspiration from generations of physicists before me who viewed their work as a means for understanding and appreciating God's creation. At the same time, I also rejected the idea, sometimes hinted at by my Christian peers, that my work was inherently less valuable than Steve's because of its secular nature. As a friend of mine once pointed out to me, Jesus himself was a craftsman by trade, and if God himself would spend part of his time on earth blessing people outside the context of formal ministry, I should feel at liberty to do the same.

My faith helped me not take my work too seriously—and not to be overly discouraged in the face of failure. As a grad student, I started taking a Sabbath day of rest every week, a practice stipulated in the Ten Commandments and observed by Jews and Christians for millennia. Every Sunday, no matter how stressed I might have felt, I took the day off from physics, a little reminder that God's love for me didn't depend on my productivity. By the time Monday rolled around, I was itching to get back to my calculations, and I saw a boost in productivity that likely more than made up for the day of rest. The understanding that my true value isn't tied to my career achievement also helped me put criticism of my work in perspective, and it made it easier for me to admit when I made a mistake (which, as my collaborators will tell you, was quite often).

29

PEOPLE OFTEN ASK ME WHAT IT'S LIKE to be a person of faith in the field of science. It's a hard question to answer, because my experiences have varied widely.

Sometimes, physicists will ridicule religion. Once, while visiting the University of Texas to give a talk on my research, I went to lunch with a number of physicists, including the late Nobel laureate (and outspoken atheist) Steven Weinberg. Unaware of my religious leanings, Weinberg began the lunch with a pointed question toward the anti-evolution movement: "Do all these people who reject evolution also reject cosmology?"

I thought about explaining the difference between young-earth creationists and old-earth creationists, but ultimately held my tongue.

Sometimes, physicists simply steer clear of religious topics. One day when I was a postdoc at the Institute for Advanced Study in Princeton, I was invited to have lunch with Ed Witten—quite possibly the greatest living theoretical physicist and maybe the smartest man alive—and with a man whose donations to the Institute helped pay my salary. A quick online search had made the donor aware of my religious views, so he spent the entire lunch asking me (very respectfully) about my opinions on religion and politics. It was probably the most stressful conversation I've ever had—talking about Jesus and Donald Trump with the smartest man alive and the man who paid my salary.

During the entire conversation, Ed Witten was surprisingly quiet. His only remark came when we were discussing God's miraculous intervention. "I think a lot of people wish God would intervene more often," he said.

Sometimes, physicists respect religion. Several of my colleagues have expressed admiration for my religious faith, or religious faith in general, though they themselves do not have any religious convictions.

Sometimes, physicists embrace religion. I don't know very many Christians in my field, but whenever I meet one, I feel an immediate kinship. Our scientific drive for knowledge pushes us to learn as much as we can about the physical universe, and that same drive pushes us, as Christians, to learn as much as we can about God. The result is a common language of science, theology, and philosophy not so different from the "twin telepathy" my brother and I have shared since childhood. Though sometimes it is discouraging that so few of my colleagues embrace religious faith, it is encouraging—perhaps even more so—that the ones who do are so strong in their faith and so capable of defending it intellectually.

In much of the world, there is intense animosity, and sometimes even violence, between people of differing religious faiths. Perhaps it's because we religious physicists represent a minority in our world, but I've certainly never felt anything like that from my Jewish, Muslim, and Hindu colleagues. And I hope they've never felt anything like that from me. Rather, there seems to be a sense of solidarity among religious scientists. Though there are important differences between our faiths, there's an even deeper sense of mutual respect among us: I've probably received more comments of admiration regarding my faith from Jewish colleagues than I have from Christian ones, and a Muslim colleague once told me that my public interviews and articles on science and God had strengthened his own faith.

On the whole, though, I can say with certainty that I have never felt persecuted or personally attacked for my faith. There are places in the world where Christians are suffering for their faith. But America is not one of those places. I can go to church, pray, read my Bible, and even write books like this one without fear of losing my job. Some of my colleagues may not agree with my faith, but fortunately my success in physics depends on my ability to do physics, not on how I worship in my free time.

Though science and faith are often viewed as enemies, I can also say I have felt less hostility toward religious faith in the upper echelons of physics than at the lower levels, or in the soft sciences or humanities. Anthropology, history, and religious studies departments are famously dismissive of Christianity—a trend many of my Christian friends and I experienced during the course of our university studies.

One of my friends who studied chemistry at Princeton had a high school science teacher who forced the class to learn the definition of a so-called scientific theory—an explanation for some natural phenomenon supported by a vast body of evidence—to refute the common creationist retort that "evolution is only a theory." But when he got to college, my friend soon realized that such definitions are nonsense: In practice, scientists use the term *theory* to describe many different things. Some theories, like quantum field theory, are among the best tested phenomena in all of science. Other theories, like string theory, lack any experimental verification whatsoever.

My high school physics teacher—who was one of the best and most important teachers I ever had—occasionally made snide remarks about religion. Yet at Cornell, Harvard, and Princeton, I met several religious physics professors. One professor even suggested to his class that God might be the best explanation after all for the fine-tuning of the universe for intelligent life—and he wasn't even a theist.

Now, it's also true that most of my extraordinarily brilliant colleagues do not embrace religion. But I've found that their reasons are generally quite ordinary. If you ask the average atheist why he or she doesn't believe in God, you'll probably get some version of the problem of evil: "If God is all-powerful, all-knowing, and all-loving, why do evil and suffering exist?" If you ask one of the world's most brilliant scientists why they don't believe in God, you'll probably hear the exact same thing.

That's not to say that the problems of evil and suffering are easy for theists to deal with. It's simply that the most brilliant minds don't have a huge advantage over others when it comes to questions of faith. We all have basically the same questions, objections, and doubts. In my experience, the ones who find answers to these questions are typically those who need answers the most. Personally, before Steve's conversion and subsequent conversations with me, I never felt much need for religion, as I was generally able to get by on my intelligence alone. Perhaps other scientists feel similarly.

Finally, I have found that most scientists—even non-religious ones—believe in some sort of power greater than ourselves. It's very common to hear physicists refer to Nature as a sort of placeholder god. For example, Ed Witten once said in an interview, "If I knew how Nature has done supersymmetry breaking, then I could tell you why humans had such trouble figuring it out." There is a widespread acknowledgment that Nature has chosen a particular way for our universe to be, and it could have chosen something different.

What's the difference between this Nature and the God (capital G) I believe in? I think the biggest difference is simply that Nature doesn't really care much about the affairs of humanity, whereas God does. Most everyone would agree that Nature has a preference for order, simplicity, and beauty, but many balk at the suggestion that it would concern itself with the affairs of one particular species on one little insignificant planet. We humans are, to quote astronomer Carl Sagan, nothing but "a mote of dust in the morning sky."[6] Why would God care about us?

To this, I like to point out that size is not a very good measure of value. I care more about the life of a baby than I do about most galaxies. I care more about the ten-nanometer transistors that make my computer work than I do about distant stars. And even as someone who studies black holes and the big bang for a living, I find nothing more incredible about the cosmos than the fact that it somehow birthed intelligent, conscious beings like us.

Ultimately, one can choose to view the size of our universe as a sign of our insignificance, or one can choose to view it as a sign of the great significance of its Creator—a Creator whose attention is not divided, who built and sustains the intricate workings of the cosmos, yet who simultaneously cares enough about humanity to become a human himself, to experience pain, suffering, and death so that we could have life.

30

DURING WINTER BREAK of my first year in grad school, I attended a conference for Christian students from all over the country. Most of what transpired was by now a familiar experience for me: We sang worship songs and listened to sermons from invited speakers. Apart from the fact that almost everyone in the audience was under twenty-five, it was essentially a glorified church service.

At one point, however, the attendees were divided into smaller breakout sessions, where we could go to shorter lectures from older, wiser Christians on a topic of our choosing—such as, "How do I discern the will of God for my life?" "How do I learn to pray better?" "How do I find the right church for me?" A number of topics looked interesting to me, but I was most intrigued by a breakout session promising "historical and scientific evidence for God."

As I anticipated, the session began with a discussion about the historical evidence for the Resurrection. The presentation was not especially thorough, but it was fine given the amount of time allotted.

But then the speaker switched to the scientific evidence for God.

What will he discuss? I wondered. *Perhaps the fine-tuning of the universe for intelligent life? The big bang theory?* After a semester of graduate studies, I knew enough about physics to know that these areas are theologically significant yet also quite subtle, and I wanted to see how a non-physicist would handle them.

Instead, to my surprise, the speaker launched into a criticism of evolutionary theory, arguing that the fossil record could not be trusted, and that the evidence pointed to the biblical account of creation. It wasn't surprising to me to encounter a Christian who didn't believe in evolution. But tucked away in my Ivy League bubble, I had forgotten how widespread this disbelief is. To many Christians—as well as many staunch atheists—evolution and Christianity are completely incompatible. Indeed, the speaker that day never even acknowledged the possibility of a third option—namely, that evolution and Christianity need not be enemies.

It's true that the opening chapters of the Bible seem to conflict with the widely accepted scientific account of creation, which holds that the universe is 14 billion years old, the earth is 4.5 billion years old, and humans evolved over the course of millions of years from apelike primates. But this conflict arises only if we assume that the opening chapters of the Bible intend to provide a scientific account of creation. I find this assumption suspect, because those chapters look nothing like what one would expect from a science textbook.

Instead, the creation account in Genesis 1 seems to be arranged thematically rather than chronologically, with domains created on days 1, 2, and 3 and the respective domains filled with their inhabitants on days 4, 5, and 6. The use of repeated phrases—"and God said," "and there was evening and there was morning," "and God saw that it was good"—make the chapter read far more like a poem or song than a scientific account.

My view is not that a straightforward reading of the early parts of Genesis is wrong, but rather that there is no such thing as a straightforward understanding of the early parts of Genesis. Any interpretation of these passages introduces complications and requires assumptions beyond what is written in the text itself. In the final analysis, the most straightforward interpretation might not be the most literal one, because these texts weren't written to answer the scientific and historical questions we're asking. Instead, they are intended primarily to tell us about the character of God: a loving, intentional Creator who is wholly distinct from the chaotic, violent, impersonal forces of creation featured in the creation myths of most religions at the time.

Modern cosmology, geology, and archaeology present difficulties for certain interpretations of Genesis. But well before modern science came along, questions of interpretation had arisen solely on the basis of the text itself.

For instance, Augustine and many of his contemporaries in the late fourth to early fifth centuries did not believe that the six days of Genesis 1 should be understood as literal, twenty-four-hour days. Instead, they argued that these "days" represent more of a literary framework. More generally, regarding interpretation of Genesis, Augustine wrote,

> In matters that are obscure and far beyond our vision, even in such as we may find treated in Holy Scripture, different interpretations are sometimes possible without prejudice to the faith we have received. In such a case, we should not rush in headlong and so firmly take our stand on one side that, if further progress in the search of truth justly undermines this position, we too fall with it. That would be to battle not for the teaching of Holy Scripture but for our own, wishing its teaching to conform to ours, whereas we ought to wish ours to conform to that of Sacred Scripture.[7]

Throughout the nineteenth century, cosmology and geology calculated the age of the universe and earth with increasing precision, and for a long time Christians more or less accepted these dates. Widespread acceptance of young-earth creationism is largely a modern phenomenon.

Personally, I never had much trouble embracing a non-literal view of the opening of Genesis. I was taught to accept evolution from a very young age, and that belief was never seriously in doubt. The bigger problem I had with evolution was its lack of efficiency: The path to humanity is very long and inefficient, involving billions of years of mutations, death, and mass extinctions before humans finally emerged. Why would God go through that whole long song and dance just to get to humankind?

But efficiency is only a concern if you're dealing with finite time and resources, and God is not. I think God is not so much an *engineer* trying to build his universe in the quickest and most cost-effective way possible, but rather an artist who builds his universe in the most beautiful way possible. Perhaps God finds beauty in evolution much like evolutionary biologists do.

Nowhere in Scripture is this clearer than chapter 38 of the book of Job. In this book, a series of great misfortunes befall Job, and he and his friends want to know why God is allowing this to happen. After thirty-seven chapters of finger-pointing from Job's friends and whining from Job, God finally responds to them out of the storm.

But he doesn't respond the way you might expect. He doesn't explain why Job is suffering. Instead, he reminds Job, "Hey, I'm God, and you're not."[8] He does this more specifically by pointing out a wide array of natural phenomena that Job doesn't understand, but that God himself made with his bare hands:

Where is the path to the source of light?
Where is the home of the east wind?

Who created a channel for the torrents of rain?
Who laid out the path for the lightning?
Who makes the rain fall on barren land,
in a desert where no one lives?
Who sends rain to satisfy the parched ground
and make the tender grass spring up?[9]

Not long ago, I saw an episode of *Our Planet* called "From Deserts to Grasslands," and one of the deserts they showed featured exactly the phenomenon described here in Job: It rains only once a decade or so; yet when it does, the desert is transformed from a barren wasteland to a place that is teeming with plant life.

In Job 38–39, God says, in effect, "I don't even care if there's no one around to observe this. Because I don't do it for you; I do it for me! I find beauty in the natural world I have created. It's not about you."

God's affections are not watered down as they are divided. He cares deeply for humankind. Yet he also cares deeply for the natural world. After all, God is not a fugitive from the laws of science, cowering in the face of scientific progress; no, he is their legislator. He built the entire universe to function with regularity, precision, and purpose, and through science he has allowed us a little glimpse into the wonder of his creation, so that we may appreciate its beauty and learn to treasure it the way he does.

31

ONE OF THE HIGHLIGHTS OF MY CAREER thus far as a physicist was getting a postdoc offer from the Institute for Advanced Study in Princeton after I finished my doctorate at Harvard. Ed Witten himself called me on my cell phone on December 7, 2016, to make the offer.

IAS, or simply The Institute, is more or less the center of the world of theoretical physics. It's where Albert Einstein worked after he came to the United States in 1933. Today, it is the premier research institution for postdocs and home to many of the most renowned theoretical physicists in the world.

I celebrated that evening with friends and champagne. At my request, Steve and his wife gave me a celebratory champagne shower, like Major League Baseball teams do whenever they clinch a trip to the playoffs. It was the culmination of years of hard work, all the sweat and tears that went into my PhD validated.

I had believed that receiving this postdoc offer would make my life complete: I had dreamed for years of obtaining this *thing*, and now I had it. What more could I want?

But climbing the mountain of success led to one of the worst emotional valleys of my life. Within a month of receiving the offer from The Institute, I found myself struggling to get out of bed, occasionally even throwing up, incapable of doing any work, severely doubting my faith in God.

What happened? There were probably a number of factors at play. One was that my life, all of a sudden, lacked direction. Of course, there was still the long-term dream of becoming a professor, but the immediate goal I had spent the previous four years obsessing over had been achieved. And instead of the long-awaited promised land I had expected it to be, I found that the success was hollow: The world continued on, just as it had before.

So I needed a new obsession, and my brain quickly latched onto my faith. Was it really true after all? I'd answered this question back in 2010, when I came to faith during my first polygraph. I had wrestled with it hard in the winter of my first year of graduate school. But all of a sudden, that wasn't good enough. I felt compelled to answer it anew every day, every hour, even every minute. Any moment in which I wasn't obsessing about my faith felt like a cop-out—like if I wasn't thinking about it, it meant that deep down I knew it wasn't true, and I was just avoiding the debate because I didn't want to face the truth.

If that doesn't sound like very rational thinking, it's because it wasn't. I soon realized that what I perceived initially as intellectual doubt was in fact a manifestation of an anxiety disorder. It wasn't the first time I had experienced the effects of anxiety, and it wouldn't be the last. Perhaps my first glimpse of the condition was the polygraph process, when I couldn't keep my brain from going into overdrive even though it meant I would fail the test.

The first time I felt like something was really wrong was during the fall semester of my senior year of college. In the months before I applied to graduate schools, stress and overwork led to consistent insomnia. It would take me two or three hours to fall asleep every night. At one point I set a curfew for myself: No homework past 10:30 p.m. Sadly, it didn't resolve my sleep issues.

But my first and last years in grad school were the worst, in part because my mental health issues came with serious doubts about my faith. In December of my first year at Harvard, I came across a blog post from a former Christian turned atheist. This blogger didn't make any compelling points, but for some reason the reminder that some people's intellectual awakenings had led them *away* from God, instead of *to* him, hit me in a way that it hadn't before. I started wondering whether my faith were mere wishful thinking, and if I, like this blogger, might someday lose my faith in God. This led me to obsessive doubts, very similar to those I would suffer four years later.

My experience of anxiety is not unusual, as close to one-third of adults in the US have some sort of anxiety disorder. During a bout of anxiety, the brain goes into fight-or-flight mode, as if there were an imminent threat that must be dealt with at all times. It didn't matter if I was sitting in class, playing intramural softball, or watching a movie with friends—my mind constantly felt as if I were in mortal peril. Some days were nonstop pain from morning until bedtime. Some days my only prayer was, "God, you're doing a pretty lousy job up there, making me go through all this."

Fortunately, no one in my Christian community ever tried to convince me that my anxiety was a sin, a sign that I wasn't trusting God enough. Some, including a counselor I consulted at my church, told me it was strictly biochemical. Religion was simply the arena of thought in which my anxiety disorder played itself out—the way some other people might experience anxiety when they think about flying or public speaking. My anxiety homed in on my religious faith not because my faith was weak but because it mattered so much to me.

Others suggested it might be a spiritual attack, a sign that "Satan is afraid of you." That possibility was actually quite flattering, but I didn't know whether I should believe it. I could understand why some people in my position would attribute their unwanted, obsessive, anxious doubts to demonic activity, like the proverbial "devil on the shoulder" whispering lies into the ear of a cartoon character. But my post-enlightenment, Western mind had difficulty with the concept of spiritual warfare and demonic attacks. I couldn't deny that such spiritual realities might exist; I just didn't know how to distinguish them in practice from ordinary mental health issues. I also wondered whether there was even a meaningful distinction between the two.

Many of my Christian friends and mentors reminded me of the importance of suffering in the life of a Christian. They encouraged me not to view suffering as something to be avoided, but as something to learn and grow from. They pointed me to verses such as 1 Peter 2:21, "God called you to do good, even if it means suffering, just as Christ suffered for you. He is your example, and you must follow in his steps."

I found refuge during that season in the music of Owl City—the one-man band Adam Young created in his parents' basement in the small town of Owatonna, Minnesota. Owl City hit it big with the 2009 song "Fireflies," which Steve introduced me to just a couple of weeks before I became a Christian. Listening to Owl City reminded me of Steve, our conversations about God and the meaning of life, and the days when I was just discovering that the world was a far more beautiful place than I had ever imagined it could be. When I was in the throes of anxiety, Adam Young's hopeful lyrics reminded me that the world is still a beautiful place.

Yet I still found myself struggling with anxiety-induced vomiting and the same recurring question about my faith: Do I really believe this stuff?

32

ONE DAY, I SCHEDULED A MEETING with my pastor to ask him that very question.

"I'm really struggling with the Old Testament," I told him. "I know some Christians feel like they have to believe every word the Bible says, but right now I'm not sure if I can believe *anything* the Old Testament says."

My pastor was a seminary professor in the field of Old Testament studies, and he told me about some archaeological discoveries that had confirmed portions of the Bible. His knowledge was impressive and his arguments were persuasive, and I left his office feeling a bit more optimistic. But by the end of my short ride home on the Boston subway, my doubts had reemerged. It seemed like for every answer

I found in response to some difficulty with the Bible, I had five more questions.

The Bible had played a pivotal role in my journey to faith. My journey began, you might say, when I started reading the New Testament that Steve gave me. It culminated right around the time I finished reading the New Testament—less than three weeks before my first polygraph.

Without the Bible, I never would have encountered the teachings of Jesus. I never would have come to believe in his miracles and resurrection. I love how different passages of the Bible, written by authors hundreds of years apart, fit together like a puzzle. As I read the Bible, I came to love the way it spoke to me, the way it convicted me of areas in which I needed to grow, yet simultaneously assured me of God's love and forgiveness.

Yet as I continued reading the Bible, I started coming across more and more passages that I didn't love—to put it lightly. I read God's stipulations for slavery in the Old Testament, wondering why he didn't just abolish the whole practice altogether. I read passages in which God instructed the Israelites to totally destroy another nation, leaving no woman or child alive. I read Old Testament laws that prescribed the death penalty for homosexuality, cursing one's parents, practicing witchcraft, and the like.

I found myself thinking about all the people throughout the centuries who were enslaved, tormented, or killed in the name of God on the basis of these biblical passages. I couldn't understand how the God of the Old Testament could allow (even encourage!) such injustices, while at the same time seem so preoccupied with cleanliness laws and purity rituals that, as far as I could tell, had little or no basis in science.

I started to research the discrepancies between the archaeological record and the Bible. There is no archaeological evidence that the Exodus ever took place. The biblical date of the destruction of Jericho is hard to square with carbon dating. A worldwide flood, like the one that supposedly hit Noah, does not fit with geological or paleontological data.

I found that Christian apologists had explanations for all these things, some of which I found compelling, some of which I didn't. For example, slavery in the Old Testament was supposedly very different from the type practiced in America—more like what we'd call indentured servitude. When God commanded the Israelites to destroy another nation, he was using Israel as an instrument of justice, and he was ensuring that the unjust practices (such as child sacrifice) of these other nations were ended once and for all.

Some of the Old Testament laws may seem harsh, but perhaps this is because we are underestimating God's holiness and the destruction that even our seemingly minor offenses wreak on the world around us. After all, my own sense of justice has been heavily influenced by modern Western culture. Why are my culture's views on right and wrong necessarily the correct ones? People in Nazi Germany, and in America during the time of slavery, somehow convinced themselves that the atrocities perpetrated right before their eyes were perfectly acceptable, their moral senses perverted by the culture around them. I'm pretty sure that twenty-first century

America isn't nearly as bad as those societies, but those historical examples nonetheless give us reason to pause and consider our own cultural biases before condemning any other code of morality.

Some of the purity and cleanliness laws may have made more sense in that culture than they do in ours. A common theme of the Bible is that we're not supposed to value luxury, comfort, money, or anything else more highly than God. What exactly counts as a "luxury" varies between cultures. Not many in our society would consider "raisin cakes" a delicacy, but from the way God complains about raisin cakes in Hosea 3, it sure sounds like they were once an unnecessary luxury that his followers were supposed to avoid. Perhaps some of the cleanliness laws are similar—certain things were forbidden not because they're inherently bad, but because the Israelites were supposed to remain separate and distinct from the wicked nations around them.

Perhaps Noah's flood was a local flood, rather than a global flood—there is evidence of major flooding in the ancient Middle East. The Exodus may not have left any archaeological evidence, but on the other hand, a bunch of nomads crossing the desert might not be expected to leave much of a trail anyway. The carbon dating of the battle of Jericho might be a bit off—after all, it's not the only historical puzzle caused by carbon dating.

A quick online search will give you plenty of answers like these to just about any objection you might have about the Bible. Some of these answers were very compelling to me, and my objections immediately dissolved. Sometimes, it took a bit more effort, but I eventually came to an explanation that satisfied me.

But other times, I found the explanations offered by Christian apologists to be very unconvincing. Some things in the Bible—especially the Old Testament—just seemed totally backwards to me. It's not just that I didn't like them personally—it's that they seemed completely out of line with what I believe is true of God's character, on the basis of Jesus' teaching.

I remembered one day when I was eleven or twelve years old, sitting in the back seat of the car with Steve as Dad drove us home from a Little League game, and the topic of the Bible somehow came up. "Do *you* believe everything that's in the Bible?" Steve asked our parents.

"Well, not *everything* in the Bible," Dad responded. He didn't elaborate on which parts of the Bible he believed and which ones he didn't, but for years afterward I assumed that Dad's primary issue with the Bible was the supernatural elements in it. But now I started to wonder whether there was more to his skepticism than that. Maybe the issue wasn't just the notion that God would speak through the pens of mortal men, but rather that the words of the Bible couldn't possibly be the words of an all-powerful, all-knowing, all-loving being.

It felt like I was having to work way too hard to try to resolve all my doubts about the Old Testament, as if my entire approach to the Bible wasn't the right one. Perhaps the Bible really was offering a unique window into the true nature of God. But if so, that window was dirty, and I couldn't tell if the God I saw on the other side was the good, loving God I trusted or someone far more sinister. C. S. Lewis, lamenting over the death of his wife, said it well: "Not that I am (I think) in much danger of ceasing to believe in God. The real danger is of coming to believe such dreadful things about Him. The conclusion I dread is not 'So there's no God after all,' but 'So this is what God's really like. Deceive yourself no longer.'"[10]

Yet if I was having so much trouble finding God in the Bible, where else was I supposed to find him?

33

IF GOD HAS THE CAPABILITY of revealing himself clearly and universally, as Christians say he does, then why hasn't he? Why do we have to work so hard to believe in this stuff in the first place? Why are debates about the existence of God even a thing?

When I give talks about the relationship between science and faith, I sometimes like to say that science does an incredible job of explaining the world around us, but atheism explains nothing. Atheism makes no predictions whatsoever for the world around us. It doesn't predict evolution, or the big bang, or even the fact that science works at all.

However, that isn't quite true. By virtue of its name, atheism does make one prediction: There is no God. This means in particular that we shouldn't experience any revelation from God. And this is why the "hiddenness of God" is a problem for theists: The one time atheism makes a prediction, it seems like a pretty good one.

On the other hand, it isn't completely obvious that God ought to show up every time we beckon him, like Aladdin's genie. After all, the God of the Bible is not a magical creature wandering about in the universe, like a unicorn or the tooth fairy. He is the creator of the entire universe, a disembodied mind who exists outside of time and space. He doesn't relate to us like a character in our story, as Harry relates to Hermione. He relates to us as the author of the story, like J. K. Rowling relates to Harry. It would be silly for Harry to conclude that there is no author to his story just because J. K. Rowling has never audibly spoken to him or appeared to him in bodily form.

So perhaps I shouldn't expect to see God in bodily form, I reasoned. But still, couldn't I expect to see clearer signs of his presence? The opening chapters of Genesis suggest that, in a perfect world, God's presence would indeed manifest itself more clearly. In the Garden of Eden, God wasn't hidden from Adam and Eve. They could talk to him, seemingly whenever they wanted, and he would respond. But after Adam and Eve ate the forbidden fruit from the tree of the knowledge of good and evil, their relationship with God was severed, so that humankind today doesn't enjoy the sort of unbroken phone line to heaven that we would have in a perfect world.

In other words, Genesis suggests that the problem of the hiddenness of God is not independent from the problem of evil and suffering. In a perfect world, God would be with us, a stay-at-home father attending to our every request. But in our broken, imperfect world, our relationship with God is also broken and imperfect.

This made sense to me because, in all honesty, the only time the hiddenness of God really bothered me was when I was suffering. "O LORD, why do you stand so far away? Why do you hide when I am in trouble?" asks the psalmist in Psalm 10:1. I didn't want God to show up just so we could play cribbage. I wanted him to actually fix things!

It seemed that all my doubts and questions ultimately pointed back to the same, simple, age-old question: Where is God in the midst of suffering?

Part IV

The King

34

IT WAS SEPTEMBER 2017. I had recently moved to Princeton to start my first postdoc at the Institute for Advanced Study, and I was visiting Japan for a physics conference in Kyoto. It was the first time I had been to Japan since I lived there at age three.

A couple of days before the conference, I stopped by my old apartment in Tokyo, which I couldn't remember but had seen in pictures. I tried to recreate each of those pictures as best I could without Steve. One on the swing set where Steve and I had swung. One in front of our old apartment building. One in Komazawa Olympic Park Stadium, where Steve and I used to ride tricycles.

But the whole time, I was haunted by my anxious doubts. It was now around nine months since they'd first hijacked my brain, and I was tired of them. I was also jet-lagged and sleep-deprived from the long flight, and that only made the anxiety worse.

The day before the conference, I was supposed to take a flight from Tokyo to Osaka and then a train to Kyoto, but a typhoon grounded all flights. Irritated, I spent much more than I would have liked for a ticket on the *Shinkansen*, also known as the bullet train.

On the train, as the darkness obscured any view of the Japanese countryside, I pulled out a book about Christianity I had been intending to read for a while. It was called *One.Life: Jesus Calls, We Follow* by Scot McKnight. By the end of the three-hour train ride, my understanding of the Christian gospel had been turned upside down.

Well before my whole conversion experience, I knew the number and the names of the Gospels in the New Testament. I'm not sure where I learned this bit of trivia, but I remember one day in high school when Dad looked up from an article he was reading to quiz me on it:

"Do you know how many Gospels there are?" Dad asked, "And do you know what their names are?"

I wasn't sure, but I took a stab at it. "Four?" I guessed. "Matthew, Mark, Luke, and John?"

"Yep, that's it!" Dad said. "According to this article, most Christians in America don't even know that. Atheists, ironically, are far more knowledgeable about religion."

So, even back in high school, I somehow knew that there were four Gospels in the Bible. What I didn't know at the time is what the word *gospel* meant. I think I thought it was just another name for one of these four books, written by one of these four guys.

Somewhere near the beginning of my exploration of Christianity, Steve taught me that *gospel* is the English translation of the Greek *evangelion*, which literally means "good news." According to Steve, the "good news" of Christianity was that Jesus died on the cross for my sins, so that through faith in him, my sins will be forgiven, and I will have eternal life. As you know, I certainly didn't believe this gospel the first time I heard it. It took months of discussions with Steve and a well-timed polygraph before I got there.

But Scot McKnight challenged me to see that the gospel of Christianity is actually even bigger and better than the way Steve first described it to me. After all, Jesus didn't define the gospel primarily in terms of forgiveness of sins, but rather in terms of the coming of the Kingdom of God: "The time promised by God has come at last! . . . The Kingdom of God is near! Repent of your sins and believe the Good News!" Jesus says in Mark 1:15. In Luke 4:43, he goes so far as to say that proclaiming this message is the reason why he was sent: "I must preach the Good News of the Kingdom of God in other towns, too, because that is why I was sent."

This is the good news according to Jesus: The Kingdom of God has come near. Okay, fair enough. But then, what is the Kingdom of God?

In the Old Testament, the people of Israel ask God to send them a king to lead them. But they soon find that all these kings are unworthy: They are corrupted by power and lead the nation of Israel astray. The people of Israel begin to dream of a king who can lead them to victory over their enemies and bring justice and peace to the land, to restore it to the way it was always meant to be.

When Jesus taught his closest followers how to pray, this is the first thing he said: "Our Father in heaven, may your name be kept holy. May your Kingdom come soon. May your will be done on earth, as it is in heaven."

Jesus didn't pray for our souls to go to heaven. He prayed for God's Kingdom to come to earth, for heaven to break into our present reality, for the wrongs of our world to be set right.

And Jesus didn't just *pray* for the Kingdom to come to earth: He enacted it; he made it happen. Modern, scientifically minded people often take issue with the idea of miracles, because we view them as violations of the laws of nature, and thus they run counter to the way the world is supposed to be. But Jesus' miracles did not represent a *breakdown* of the laws of nature—a *breakdown* of the way the world should be—but rather a *restoration* of the world to the way it should be. Jesus' miracles were, to him, instances of the coming Kingdom of God, instances of heaven breaking into our world. The sick were healed, the blind could see, the hungry were fed, the dead were raised to life, and death itself was conquered in Jesus' crucifixion and resurrection. Jesus viewed his atoning death on the cross as the archetypal event of the Kingdom of God, an example of sacrificial love for his followers to emulate.

The shocking claim of Christians throughout the ages is that Jesus is the long-awaited King of kings. This claim is surprising because, in many ways, Jesus looks nothing like the king that people expected. He came not with a sword, but with a healing hand. He came not to defeat the Israelites' Roman oppressors, but rather to defeat the greater human enemies of sin and death. He is a king with a cross for a throne. He is a king with a crown made of thorns. He is, in many ways, the opposite of every human king.

Likewise, his Kingdom is, in many ways, the opposite of every human kingdom: "Whoever wants to be a leader among you must be your servant, and whoever wants to be first among you must be the slave of everyone else. For even the Son of Man came not to be served but to serve others and to give his life as a ransom for many."[1]

McKnight's book helped me understand that the world actually matters to God. God cares about my work and my anxiety. He cares about poverty, injustice, and the broken systems in our world. Through the work of Jesus, God's great project to redeem and restore his broken world has already begun. And ultimately this project will succeed because Jesus is King.

Years before, I had come to Jesus in search of a savior, recognizing for the first time my moral imperfection and need for forgiveness. But to get through my season of anxious doubts, I needed a king powerful enough to bring an end to my suffering, humble enough to stand with me through it, and great enough to be worth suffering for.

35

ONE DOUBT-FILLED JANUARY DAY, I found myself in tears on the floor, curled up in a fetal position. Realizing I needed help, I called a friend of mine from college.

"The Bible says that once we come to Jesus, he'll never drive us away," I said. "Yet we both know plenty of people who follow Jesus for a while, then stop. So how do I know that I'm not one of them?" I asked. "How do I know that I'm not going to lose my faith someday?"

"Because you don't want to," my friend responded. It was simple, but profound. The fact that the thought of losing my faith bothered me so much was a good sign. If I wanted so badly to emerge from this time of doubt as a Christian,

204 ≈ CHASING PROOF, FINDING FAITH

I probably would. Kind of like Harry Potter begging the Sorting Hat not to put him in Slytherin, I had agency in where I would end up.

But this raised another issue: Had I become a Christian because I *actually* believed it was true, or because I *wanted* it to be true?

This is a common criticism of Pascal's famous "wager." Pascal argued that if Christianity is true, the value of believing in it is infinite: an eternity of joy. If Christianity isn't true, then the cost of believing in it is finite. The expected value of believing in Christianity is therefore infinite, so it is better to believe in it, if you can.

Lots of people have pointed out the obvious flaw in this logic: Is believing really a choice? I couldn't believe that the earth is flat, even if you offered me a million dollars. I could *say* I believed that, but deep down, I'd know I was lying.

This was something that really bothered me when I first struggled with doubt. All of a sudden, I didn't know if Christianity was true. Did that mean I wasn't really a Christian? How confident did I have to be in order to call myself a Christian: 51 percent? 90 percent? 100 percent?

Some Christians will say you have to be 100 percent sure. I think that's nonsense. I'm not 100 percent sure that the sun will come up tomorrow. I'm not 100 percent sure I'm a person rather than a bat dreaming he's a person. I'm not 100 percent sure of anything.

When I finished reading *One.Life*, I decided to check out some of the other books Scot McKnight had written. After a bit of online searching, I came across one with an especially intriguing title: *Salvation by Allegiance Alone*—written by Matthew W. Bates with a foreword by Scot McKnight.

Bates emphasizes that faith in the sense of "intellectual belief" is not really what God wants from us. After all, as James points out in James 2:19, "Even the demons believe." Bates explains that the Greek word *pistis*, which is typically translated as "faith" or "belief" throughout the New Testament, can also be translated as "allegiance" or "loyalty" (and in some non-biblical sources produced around the same time as the New Testament, these latter translations are undoubtedly the correct ones). Thus, what God really wants from us is not intellectual belief that he exists, but rather unswerving allegiance to Jesus the King.

Ironically, I think my education kept me from understanding this sooner. In my social circles, God is so often viewed as an intellectual *question* to solve rather than a *person* to love, worship, and obey. But in my own experience, though the intellectual arguments gave me compelling reasons to believe, they never really gave me peace. The day I became a Christian wasn't the day I decided Christianity was more likely true than false: It was the day God reached out to me and I decided to put Jesus on the throne of my life and give him my allegiance.

Allegiance is easier to control than intellectual belief. I may not be able to push a button and switch off the doubts, but I can choose to take the side of God against my doubts.

C. S. Lewis expresses this well in his book *The Screwtape Letters.* The book was a bit confusing at first because it's written from the perspective of an older demon named Screwtape counseling a younger demon named Wormwood on how to be evil and ruin a human's relationship with "the Enemy" (i.e., God). So whatever advice we read in the book, we're supposed to do the opposite as Christians. Screwtape writes:

> Do not be deceived, Wormwood. Our cause is never more in danger than when a human, no longer desiring, but still intending, to do our Enemy's will, looks round upon a universe from which every trace of Him seems to have vanished, and asks why he has been forsaken, and still obeys.[2]

I am convinced that this is what God wants from me: not that I be free of doubts, but that I continue to obey him even in the midst of my doubts. After all, doubts come and go—some days I may feel confident that Jesus rose from the dead; other days, I may wonder whether it's all nonsense. But even on those days, I can *choose* to go to church, to pray, to meditate on the life and teachings of Jesus and to try to put them into practice.

Of course, this isn't exactly a walk in the park. I remember how difficult it was trying to lead a Bible study during those seasons of anxious doubt, trying to talk about my faith with others, preparing talks on science and religion, and feeling like a hypocrite. I was trying to convince other people to adopt a faith that wasn't like the one I was experiencing. I wanted them to find hope and joy and peace, but all I was experiencing was doubt and pain and anxiety and fear that it wasn't true.

Yet my faithfulness in those seasons wasn't wasted. So many of the highlights of my Christian journey were forged in the furnace of doubt, fear, and anxiety, as I learned how to walk alongside others in their own doubt, fear, and anxiety.

During one of those times when I was overcome with doubt, Steve pointed me to John 6:67-68. Jesus has just preached a tough message that the crowds of people following him don't like, and as a result they begin to disperse. When the dust settles, just Peter and the other disciples are left.

"Then Jesus turned to the Twelve and asked, 'Are you also going to leave?' Simon Peter replied, 'Lord, to whom would we go? You have the words that give eternal life.'"

As I read Peter's response, I always get the feeling that he had indeed considered the possibility of leaving Jesus. And can you blame him? Jesus' strange teaching—"My flesh is true food, and my blood is true drink"[3]—has convinced the rest of the people that he is off his rocker. At this point in the narrative, Peter has no idea that Jesus will someday rise victoriously from the dead in vindication of everything he has taught them. At this point, Peter would have been well-justified in doubting Jesus' words, in looking for a safer option, in leaving Jesus altogether like so many others.

Yet Peter decides to stay. And from his initial response, "Lord, to whom would we go?" we can sort of understand his thought process: He has considered his other options. He has weighed the pros and cons. But when he sums it all up, he realizes that his best option is this strange, enigmatic, miracle-working preacher standing in front of him. Jesus' words may not lead to eternal life after all. But nothing else will either.

I am not 100 percent certain that Christianity is true. Maybe all the evidence, everything I've seen and experienced that has brought me to God, has been nothing but a mist. Maybe there's no deep meaning to this life after all, and someday I'll die, and that will be the end of it.

However, I am absolutely convinced that the best chance I have of making it out of this world alive is through the Cross of Christ.

Ultimately, the question is not whether or not we should have faith. The question is *where* should we place our faith? The opposite of faith in God is not doubt, or certainty, or science—the opposite of faith in God is faith in something else. And as Peter realized that day, nothing else will do. Everything else is a broken cistern, a vessel that may hold our hopes and dreams and affections and give us pleasure for a little while, but will ultimately leave us thirsty and wanting more than it can deliver. Placing our lives in Jesus' hands is a big risk, but the alternatives are worse.

And risky though it may be, it's worth it. Think, after all, of all that people sacrifice in pursuit of their dreams—of stardom, of fortune and fame, of success in music and athletics and theater and tech. In my own field of string theory, dozens of starry-eyed graduate students show up every year at universities across America in hopes of becoming the next Albert Einstein or Ed Witten, but only a small handful will even manage to land faculty jobs in the field. Very few of us dreamers will actually make it big, but our dreams are big enough to be worth the risk for us.

In Christ, we have reason to dream much bigger dreams: dreams of eternal life, joy, peace, and unconditional love. If anything is worth sacrificing for, it's this.

36

JESUS WAS CALLING ME TO ALLEGIANCE, even in the midst of my doubts. But what did that actually look like? What did God want from me?

In May 2020, I was in Covid quarantine with relatives just outside of Minneapolis. One afternoon, I took a break from physics to get a glass of water and found my cousin Alex in the kitchen looking distraught.

"A Black man here in Minneapolis just died after a police officer knelt on his neck," she said.

"Oh my gosh," I said. "Right here in Minneapolis?"

In the coming days, we learned more about the murder of George Floyd, who died lying in the street in handcuffs, gasping for breath, while a Minneapolis police officer knelt on his neck for more than nine minutes. Within hours, the city was in an uproar.

Like so many others, I wanted to help. But in practice, what could I actually do? All I could think of was to tag along with Alex as she headed downtown, and together we joined a group of protestors who were cleaning up a Target store that had been destroyed in the aftermath of Floyd's death.

Afterward, we joined in one of the protests, trying our best to stay socially distanced at the back of the crowd to avoid contracting Covid-19 while the leaders at the front led chants like "No justice, no peace!" and "Two, four, six, eight, America was never great!"

At one point, a young woman got hold of the mic, and quite unexpectedly—to me at least—gave a little sermon. As a young man next to her held up a sign with Bible verses about love and forgiveness, she read John 2:19-21: "Jesus answered them, 'Destroy this temple, and I will raise it again in three days.' They replied, 'It has taken forty-six years to build this temple, and you are going to raise it in three days?' But the temple he had spoken of was his body."[4]

"That's us!" she triumphantly concluded, to great applause from the crowd. "We are the body!" I recognized it immediately as an allusion to 1 Corinthians 12, where Paul declares that the church—the collection of Jesus' followers worldwide—is the body of Christ. In other words, the church is God's "hands and feet" in the world, working together to bring about his Kingdom on earth, with each part playing its own distinct role in the team effort. This young woman was playing her role that day, and she was playing it to perfection. She had seen a broken part of our world and was committed to mending it, driven by the vision of God's Kingdom she had found in the pages of Scripture, and fearlessly willing to tell others about her faith along the way.

At that moment, I realized this is the sort of church Jesus had in mind—a ragtag group of people from all walks of life, all races and ethnicities, all socioeconomic classes, bonded together as one in pursuit of justice and equality for all. The sort of allegiance Jesus demanded from me was not merely the practice of certain spiritual routines, but rather wholehearted commitment to everything—and everyone—that Jesus himself stood for.

Back in the summer of 2009 when I first started reading the New Testament, one of the things that most drew me to Jesus was the way he embraced the perspective and the cause of the powerless, the way he spent his time with the people on the fringes of society. As I reflected on the years since then, I realized that the times I felt closest to God were the times I followed Jesus' example.

I'm thankful that my faith has provided me with a diverse community of friends from all different backgrounds, many of whom I never would have encountered within the ivory tower of academia. At the same time, I regret that I have so often gravitated toward highly-educated circles of society, which has distorted my view of the world and driven my impostor syndrome and anxiety. I can tell that the barrier-breaking way of Jesus is a better way, not only for the sake of the poor and powerless, but for my own sake as well.

37

IN JANUARY 2021, AN ATHEIST YOUTUBER named Tom Jump invited me to join him the following month for a live, online discussion about the existence of God. At first, I thought it was a joke, but a quick internet search revealed that Tom was in fact a rather prolific YouTuber, and he had previously hosted a number of prominent theistic thinkers. I accepted Tom's offer almost immediately.

I had had many informal conversations about the existence of God with atheists before, and I had given many talks on miracles and the dialogue between science and faith, but this was the first time I had been invited to a conversation with a professional atheist. I was excited, but also so nervous that I couldn't sleep the night I got the invitation.

I decided to combat my nerves with my usual approach to test-day anxiety: over-preparation. I recruited a small army of friends to help watch Tom Jump's past discussions and debates, like a football team watching game film of its next opponent. My friend Josh, who finished his PhD in astrophysics at Harvard the same year I finished mine in theoretical physics, joined me on Zoom for a couple of Tom Jump's debates on the topic of cosmological fine-tuning, assuming that my own discussion with Tom would probably go along similar lines. My friend Matt, who was a PhD student at Princeton when I was a postdoc there, did his own deep dive into Tom Jump's YouTube channel and ended up playing the role of "TJump" for a practice debate with me.

After watching a number of Tom's videos, we came to the conclusion that it would be hard to keep the discussion from getting stuck in the weeds. Tom's previous conversations with philosophers, physicists, and other scholars often got bogged down on some esoteric point in probability theory or philosophy. His debates with more aggressive Christian apologists routinely ended up as shouting matches, or "dumpster fires," as he called them.

"Whatever I do," I told Matt, Josh, and Steve after our practice debate, "I better not start yelling at TJump." We joked about this for a while, imagining worst-case scenarios for the upcoming discussion, as well as possible puns I could make around the fact that we were both named Tom.

Initially, I imagined I would be able to outdebate Tom Jump, dominating his atheistic arguments with my own arguments for God, leaving him and his atheistic fan base groping for words. But as the day of our discussion neared, I began to wonder whether gearing up for a debate was really the right way to approach the conversation. After all, Steve hadn't come to me looking for a debate the first time we talked about God—the beginning of my own journey to faith. And Tom Jump knew far more about theology and philosophy than I had when Steve first approached me.

Plus, as I watched Tom's videos, I became curious to know more of his story. So when our discussion began, I asked him to share more about his background.

"I was wondering if maybe, before we get started, I could ask you a few questions about your experience, your story, how you got into this, if that would be all right with you?" I asked nervously. I didn't know what I would say next if he declined.

Fortunately, Tom was more than happy to share. In many ways, his story was the mirror opposite of mine: He grew up in a devout religious family, but he suffered from severe depression—far worse than my own anxiety. He prayed fervently that God would heal him of his depression, but nothing ever changed. Eventually, Tom lost hope that there could be an all-good, all-powerful God, and he was converted to atheism by the arguments of new atheists like Richard Dawkins, Sam Harris, and others.[5]

After Tom shared his story, we got down to business. I laid out my argument for the existence of God, which of course he didn't find convincing. Tom went off script and took the argument in a direction I didn't expect, and as my nerves got the best of me, I missed an opportunity to point out a serious flaw in his argument. Tom made some good points, too, but eventually we got stuck on the subtleties of Bayesian probability theory, and most of the audience got bored or lost.

I left my discussion with Tom Jump disheartened that I hadn't performed better. But most of my friends who watched the debate were much more impressed by the opening dialogue. They were moved by Tom's story and hoped he would someday be healed of his depression.

Within a few days, most everyone had forgotten about the discussion. Tom moved on to other debates, and my friends who had helped me prepare for the debate went back to their daily lives.

But I, yet again, found myself curled up on the floor, my mind stuck in its usual cycle of anxiety, fear, and doubt.

38

I COULDN'T FIGURE OUT WHY precisely the doubts came on like they did. It wasn't as if Tom Jump had come up with some knockdown argument for atheism that I hadn't thought of before. Indeed, I wasn't even doubting the existence of God as much as I was doubting my own ability to reason about the existence of God. Maybe God doesn't exist after all, and I'm a fool for thinking he did? Or maybe God does exist, and I'm an idiot for even questioning myself?

Tom Jump's story resonated with me in a way it hadn't before. His own unending battle with depression had led him to—at least for the time being—abandon his belief in God. My own anxiety had led me to . . . where exactly? I wasn't about to abandon my allegiance to God, but I didn't know what my faith would look like when—if—I finally got better.

As a scientist, I took pride in my ability to reason in the face of uncertainty. But this sort of uncertainty was a different animal entirely. The experience of pain—and the uncertainty over when that pain might come to an end—was nothing my PhD in theoretical physics could help me with. My ability to think rationally went out the window, and my anxious mind would take a molehill of doubt and turn it into a mountain.

Caught as I was in the throes of yet another season of anxiety, my debate prep team became my support group. Some of them listened patiently and tried to reason through my doubts with me, while others encouraged me to take a break from the doubts entirely. One friend emailed me: "Dude, if you are literally dry heaving with anxiety, then that is a psychological problem, not a philosophical problem. You need *treatment*, not *arguments*."

Some of my friends reminded me that my doubts weren't unique to my particular faith tradition: If I were an atheist, I would be stuck with similar but opposite doubts about my belief in atheism, and I would find something else to be anxious about.

Many of my Christian friends had also experienced mental health issues, yet it seemed to me they were better able to cope with them than I was. They didn't start questioning the existence of God—or at least, their ability to deduce the existence of God—every time their brains malfunctioned.

"Well, that's true," said Josh when I pointed this out to him. "I don't suffer like you do over these sorts of theological questions. But the problem of suffering is the reason I walked away from God in college. And it's also what brought me back."

I knew that Josh had become an atheist his freshman year and had returned to the faith a while later. But I'd never actually bothered to ask what had led him down those opposite paths.

Josh told me that he had grown up in a strong Christian family, the son of two Indian immigrants. But by the time he started college at MIT, he couldn't make sense of the suffering in the world. Like many of his classmates, Josh had a strong desire to change the world. He could see that the world was terribly messed up, in need of deep structural changes, and he wanted to be a part of the solution. To just say "God is good" and rejoice that he'd someday go to heaven while others were suffering seemed tone-deaf at best and despicable at worst.

Josh started looking into other philosophies. He especially enjoyed reading nihilists, like Nietzsche, whom he considered incredibly brave. These people dared to explore what it meant to get rid of God, to get rid of all the mythology and deal with just the brute facts of life.

But the more Josh came to learn about nihilism, the more he came to recognize the shortcomings of his newfound belief system. These issues became especially clear when he read an article by a humanist reporter who had worked in Rwanda right after the genocide there.

"It was obvious to her that what had happened was evil. She had met all these victims of rape, murder, and torture.

Yet, intellectually, she had no right to say it was wrong. As a modern, Western, post-Enlightenment woman, all she could say was that her *culture* would say it was wrong. This was very disappointing for her, because she could see it wasn't enough."

The problems of Josh's nihilistic worldview started to have real-life consequences. At MIT, he was surrounded by privilege everywhere he looked. People had the resources to make a real difference in the world, yet everyone was so self-centered—including Josh himself.

"People were trying to prove how smart they were, how cool they were. They were out getting hammered every weekend. I couldn't stand it. And yet I was doing the same thing. So I couldn't stand *myself*. I was like, 'There has to be more.' I felt we were obliged to do more, morally, than what we were doing."

Josh had walked away from Christianity in the first place because it seemed too simplistic to deal with the harsh realities of the world. But at MIT he was finding that atheism suffered the same problem.

"There was no way to escape that place where ethics drop out from under you, and it becomes just a question of 'What do I think? How do I feel?' Why sacrifice your life for some greater calling when there is no greater calling?"

The deep drive in our culture—and in Josh's heart—to make the world a better place simply didn't fit with atheism. The tables began to turn; Josh began to wonder whether maybe there was something to religion.

Still, he didn't want to come back to Christianity, which

he considered lame and uncool. It's what he'd done as a kid, and he felt as if he'd outgrown it.

Yet as he explored the various faiths, he felt himself irresistibly pulled back to Christianity. Every other religion Josh looked at told him basically the same message he had been internalizing his entire life: You succeed in life by being the best at x. When he was growing up, x equaled math, science, chess, etc. According to the other religions, x equaled living morally, achieving enlightenment, or something else.

"Christianity was just so different, because it said that the ones who are counted righteous are the powerless, the people who know they can't do it on their own."

Christianity doesn't shy away from the problem of evil and suffering. The entire biblical narrative revolves around it—from Genesis 3, where Adam and Eve's disobedience lead to evil and suffering in the world; to the Psalms, which ask, "How long, O Lord? Will you forget me forever?"; to the book of Job, where Job and his friends stand around wondering why all of these tragedies are befalling Job; to the New Testament, where Jesus weeps at the funeral of his friend Lazarus, and the apostle John envisions a future world where every tear has been wiped away from every eye.

The final straw for Josh came when he went to a talk by pastor Timothy Keller, who visited MIT to give a talk about the problem of evil and suffering.[6] Josh wasn't invited by anyone, and he didn't have any friends who were interested in learning more about Christianity, so he went alone.

At the outset, Keller talked about philosophical responses to the problem of evil and suffering, such as the classic "free will" defense. Like many—including me—who encounter the free will defense, Josh wasn't all that convinced: Could free will really be worth all the pain and evil it has wrought?

But then, Keller changed gears: "Let's move beyond the philosophy. Let's talk about the people here who are suffering and asking, 'How could God be in this?'"

Christianity's greatest response to the problem of evil and suffering isn't a philosophical argument: It's a person. God, in Jesus, has stepped into the hopelessness of the human situation, and he has wept, bled, suffered, and died with us, and for us, so that we could have life. John Stott said it like this:

> I could never myself believe in God, if it were
> not for the cross. . . . In the real world of pain,
> how could one worship a God who was immune
> to it? . . . He laid aside his immunity to pain. He
> entered our world of flesh and blood, tears and
> death. He suffered for us. Our sufferings become
> more manageable in the light of his. There is still
> a question mark against human suffering, but over
> it we boldly stamp another mark, the cross which
> symbolizes divine suffering. "The cross of Christ
> . . . is God's only self-justification in such a world"
> as ours.[7]

Yet by itself, that isn't quite good enough. The fact that God can empathize with us in our sufferings may boost our morale, but it doesn't erase the problems with our world. Ultimately, there are some things that just have to be fixed.

There are some things that can only be solved by an undoing,
a reversal, a resurrection.

Keller ended with a quote from Dostoevsky's *The Brothers
Karamazov*:

> I believe like a child that suffering will be healed
> and made up for, that all the humiliating absurdity
> of human contradictions will vanish like a pitiful
> mirage, . . . that in the world's finale, at the moment
> of eternal harmony, something so precious will
> come to pass that it will suffice for all hearts, for the
> comforting of all resentments, for the atonement of
> all the crimes of humanity, for all the blood they've
> shed; that it will make it not only possible to forgive
> but to justify all that has happened.[8]

Josh had already seen plenty of times in his life when God
had turned his suffering around and used it for good. Keller
challenged him to ask, *What if that could be true of all of history? What if in the end, all the wrongs could be set right? What
if everything sad could come untrue?*

Josh decided that day to become a Christian again. I met
him a few years later, when we were both grad students at
Harvard. We bonded over science, lifting weights, anxiety,
and our shared faith in Jesus.

Josh helped me to see that it was too soon to judge
the goodness of the human story—or our own individual
stories—from the perspective of someone stuck in the middle
of it all. I don't know how my life will turn out, or how my

anxiety will be healed and made up for. But I can turn to Jesus—the God who himself experienced suffering—and believe with bold, reasoned confidence that God cares about suffering, is with me in my suffering, and will someday put an end to all suffering.

Already, as I looked back, I could see rare glimpses into the way my own suffering might somehow be part of a greater happily-ever-after, with my tears giving way to some of my most cherished memories of love and affection. I thought about my Uncle Jim's funeral, where I stood sobbing in front of his coffin until my grandma—Jim's mother—put her arm around me and said, "It's okay to have a good cry."

I thought back to the polygraph process, where hours and hours of painful self-reflection ultimately led to one of the most joyous transformations of my life. I thought about the rare moments of affection my family and I had shared since then, and that I was finally able to say "I love you" to them. I thought of all the close bonds of friendship I had forged throughout my journey as a young Christian, none of which would have been possible without the difficult journey that brought me to faith in the first place.

I thought back to the first time I experienced anxious doubts, when Mom overheard my sobs and knocked on my bedroom door to ask what was the matter.

Weeping bitterly, I told her I had been having a lot of doubts about my faith.

"Aww, honey," she replied. "I know it's hard sometimes. But it's worth it for the good times."

And she gave me a hug.

39

WITHIN A FEW WEEKS OF MY DISCUSSION with Tom Jump, my anxiety once again retreated—and much more quickly than usual. Again my faith survived, and the doubts that had bothered me so much seemed foolish in retrospect.

That's not to say *all* my anxious doubts were foolish. Some were, of course, but some were just emotional over-reactions to legitimate questions about the existence of God. Even after my anxiety had subsided, I often found myself pondering the hiddenness of God and the more troubling passages of the Bible, wondering whether these difficulties were destined to bother me forever. But as I continued along the road of faith, I found that answers—or at least, partial answers—would occasionally come to me.

Not long ago, a friend pointed me to a passage in a book called *Jesus Without Religion* that altered the way I think about the hiddenness of God. The book's author, Rick James, recounts a story from an annual report of the *JESUS* film project, an organization that put the Gospel of Luke on film, translated it into who knows how many languages, and showed it all over the world to people who may never have even heard of Jesus. According to the annual report, an Indonesian woman named Mrs. Peni had been blind for four years, but when she heard that in the *JESUS* film Jesus heals a blind man, she asked her youngest daughter to guide her to a showing of the film so she could hear it, even if she couldn't see the pictures.

When the scene in the film came where Jesus heals the blind man, Mrs. Peni exclaimed, "I want to see too!" A short while later, as Jesus was being nailed to the cross, Mrs. Peni's vision was restored. When the town got word of this story, and Mrs. Peni demonstrated her newfound sight at a town meeting, many people decided to become Christians.

This story is remarkable, but what was even more remarkable was a follow-up interview Rick James did with a member of the JESUS film's leadership team. When James pointed out that the JESUS film project used to publish miraculous accounts like these all the time, but no longer did so, and he wondered why, the leader replied, "We don't share those stories as much in the US, because where there is no faith, miracles actually produce skepticism. It does the exact opposite of what it was intended to do: encourage belief."

This story hit me hard, because I'm in the same boat as the average American, if not worse. When I hear miracle stories like that of Mrs. Peni, my scientifically trained mind kicks into gear, trying to come up with an alternate, naturalistic explanation: Did this story really happen? Was Mrs. Peni really blind to begin with? Was her "cure" strictly psychosomatic?

Now, I can't *prove* beyond a reasonable doubt that these alternate explanations are wrong. But I also can't prove that Mrs. Peni's healing *wasn't* actually a miracle. If Christianity is true, it wouldn't be unreasonable to think that miracles might occur every so often when people are introduced to the good news of Jesus' Kingdom for the first time.

In Matthew 12, a group of Pharisees have heard about Jesus' miraculous signs and want to see one for themselves. But Jesus responds, "Only an evil, adulterous generation would demand a miraculous sign; but the only sign I will give them is the sign of the prophet Jonah."[9] Jesus' refusal to give them a sign is surprising, because already in the chapter he has healed a man with a withered hand and a blind man who is possessed by a demon and can't speak. Elsewhere in the Gospels, he reportedly feeds five thousand people with five loaves of bread and two fish, raises the dead, cleanses the lepers, and so on. Marcus Borg, a scholar of early Christianity and the historical Jesus, remarks that "more healing stories are told about him than about any other figure in the Jewish tradition."[10]

Yet when these religious leaders come to him, Jesus sends them away empty-handed: No miracle for you, just a cryptic comment about the sign of Jonah. Why? The key difference is their approach to Jesus: Whereas some people came to Jesus in desperation, sick of being sick, wanting to get better, and were met with healing, the Pharisees came as skeptics, to test Jesus, and were left hanging.

You might *say* that a miracle will prove my authority, says Jesus. But let's be honest: That's not really what you want. Because if I really am the Messiah, the Son of God come to bring judgment to the wicked, you are in big trouble. So, though you ask me for a miraculous sign, what you really want is to see me try—and fail—to produce such a sign.

We like to think that a miraculous sign from God would quench our thirst for proof and calm our anxieties about God's seeming absence. Yet many people who saw Jesus' miracles, and even those who *believed* them, refused to follow him. Instead, they reasoned that his miraculous powers must be demonic in nature.[11] They saw God in human flesh doing bona fide miracles before their very eyes, yet they still found a way to justify their disbelief by appealing to their deeply entrenched worldview.

Perhaps we moderns aren't so different from the ancients. Perhaps miracles like these happen every day, but here in the West we are too blinded by our own functional atheism to recognize them for what they are. Perhaps the real issue isn't God's hiddenness. Perhaps the real issue is our skepticism. Maybe God isn't hiding from us; maybe we're hiding from God.

"Now we see things imperfectly, like puzzling reflections in a mirror," writes the apostle Paul. "But then we will see everything with perfect clarity."[12] Someday, God will relate to us as a husband relates to his wife, as a mother to her baby. When the world is finally set right, our relationship with God will be completely healed as well. But in the meantime, in our broken, imperfect world, our relationship with God is also broken and imperfect. And just as we should not expect to see perfection in our world before the story has ended, neither should we expect an end to the hiddenness of God before the story has ended.

God is not in our debt: He doesn't owe us a better world, nor does he owe us proof of his existence. He doesn't need to show us miracles; he doesn't need to appear to us in bodily form. Yet Christianity claims that God has done exactly that: He showed up in bodily form, in the person of Jesus. He has done miracles—healing the blind, the disabled, the sick, making our world into a better world. God has already given us more than we deserve. And the best is yet to come.

40

AT SOME POINT, I BEGAN TO REALIZE that the most satisfying answers to my doubts came not from wild philosophical speculation about the nature of God, but from a thorough look at the figure of Jesus. No philosophical response to the problem of evil and suffering was nearly as compelling to me as a God who experienced evil and suffering firsthand. My own attempts to square God's love with God's judgment did less for me than the image of a God who, in Luke 19:41-44, weeps at the thought of bringing judgment upon his city. And my faith has never been put in serious jeopardy by the hypocrisy of other Christians, because I know that Jesus had just as much trouble with self-righteous, greedy, corrupt, hypocritical religious types as I do.

After failing for years to make sense of some of the most difficult, confusing passages of the Old Testament, I eventually reached the conclusion that my approach to the Bible was falling victim to the same problem: I was trying to read the Old Testament from the perspective of a twenty-first-century scientist, when I should have been reading it through the eyes of a first-century Jew named Jesus.

There is no denying that Jesus treated the Scriptures (more specifically, the Old Testament Scriptures—the only ones around at the time) with great reverence. Like other Jewish teachers of his day, he quoted Scripture nonstop, and he gave it authority over his own actions.

Yet Jesus also held a nuanced view of Scripture. He often took issue with commandments and laws in the Old Testament:

"You have heard the law that says the punishment must match the injury: 'An eye for an eye, and a tooth for a tooth.' But I say, do not resist an evil person!"[13]

"You have heard the law that says, 'Love your neighbor' and hate your enemy. But I say, love your enemies! Pray for those who persecute you!"[14]

When the Pharisees asked Jesus if they were allowed to get divorced, pointing out that Moses permitted such a thing in the Torah, Jesus responded: "Moses permitted divorce only as a concession to your hard hearts, but it was not what God had originally intended. And I tell you this, whoever divorces his wife and marries someone else commits adultery—unless his wife has been unfaithful."[15]

These statements from Jesus would have shocked his

listeners: He, a faithful Jew, was asserting that his authority exceeded that of Moses and the other Old Testament writers. Jewish scholar Jacob Neusner has this to say:

> Here is a Torah-teacher who says in his own name what the Torah says in God's name. . . .
>
> For what kind of torah is it that improves upon the teachings of the Torah without acknowledging the source—and it is God who is the Source—of those teachings? I am troubled not so much by the message, though I might take exception to this or that, as I am by the messenger. . . . Sages, we saw, say things in their own names, but without claiming to improve upon the Torah. The prophet, Moses, speaks not in his own name but in God's name, saying what God has told him to say. Jesus speaks not as a sage nor as a prophet. . . .
>
> So we find ourselves . . . with the difficulty of making sense, within the framework of the Torah, of a teacher who stands apart from, perhaps above, the Torah. . . . We now recognize that at issue is the figure of Jesus, not the teachings at all.[16]

As twenty-first-century readers, we often want to approach the Bible as a newspaper article or a modern legal document: literal, clear, and straightforward. But the Bible is far more complex than that. It is not a single book, but a collection of many books of many different genres written by many authors across many centuries. Some of it is historical narrative—Luke emphasizes at the beginning of his Gospel that he consulted with eyewitnesses and carried out a careful investigation before writing his account of Jesus' life. But much of the Bible is written in other genres, including poetry, metaphor, allegory, and symbolism.

This is important, because we cannot truly understand the Bible until we understand how God intends for us to read it—which may not always agree with our modern expectations for the text. I think that Rachel Held Evans—a woman whose doubts may have surpassed even mine—says it well in her book *Inspired*:

> The problem isn't that liberal scholars are imposing novel interpretations on our sacred texts; the problem is that over time we've been conditioned to deny our instincts about what kinds of stories we're reading when those stories are found in the Bible. . . . That's because there's a curious but popular notion circulating around the church these days that says God would never stoop to using ancient genre categories to communicate. Speaking to ancient people using their own language, literary structures, and cosmological assumptions would be beneath God, it is said, for only our modern categories of science and history can convey the truth in any meaningful way.
>
> It is no more beneath God to speak to us using poetry, proverb, letters, and legend than it is for a

mother to read storybooks to her daughter at bedtime.
This is who God is. This is what God does.[17]

When I say that I think Genesis 1 is meant to be understood as a poem or song rather than a scientific account of
creation, creationists and atheists alike often respond with a
valid question: If parts of the Bible are meant to be read sym-
bolically rather than literally, how are we supposed to know
which is which? How do I know where allegory ends and
history begins? The truth is that I'm not exactly sure in many
cases, but I also don't think it's necessarily a problem. In some
ways, it would be nice if the Bible were simpler to understand;
it would be easier if God had just given us a philosophy textbook to tell us all the truths we would like to know about our
universe. But on the other hand, I think there's something
uniquely beautiful about the Bible we *do* have, a collection of
sixty-six books that speak with the voices of imperfect kings,
teachers, peasants, and priests throughout the ages, attesting
to the faithfulness of an almighty yet mysterious God. One
could always try to translate the Bible into a more straightforward form, condensing it into a simple list of philosophical
propositions and historical facts, but so much of the beauty
would be lost in the process. Rachel Held Evans again:

When you stop trying to force the Bible to be
something it's not—static, perspicacious, certain,
absolute—then you're free to revel in what it is:
living, breathing, confounding, surprising, and yes,

perhaps even magic. The ancient rabbis likened Scripture to a palace, alive and bustling, full of grand halls, banquet rooms, secret passages, and locked doors.[18]

Jesus often spoke in beautiful, confounding, mysterious parables. We shouldn't be surprised to find such mystery in the rest of Scripture as well.

What does this mean, practically speaking? First off, it doesn't mean I feel at liberty to simply ignore anything in the Bible that bothers me. Sometimes, the process of wrestling with the Bible leads me to a better understanding of who God is. Second, it doesn't mean I'm indifferent about the historical accuracy of the Bible, especially when it comes to the life, death, and resurrection of Jesus. If I came to believe that Jesus' resurrection was pure mythology, I would find a different religion.

But it does mean I approach the Bible with humility, knowing that two reasonable people will sometimes have very different interpretations of the text, and my first instinct isn't always the right one. It also means that before I approach the Bible for answers, I better know how to ask it the right questions. For example, as we saw in Chapter 30, difficulties arise when one tries to interpret the opening chapters of Genesis in a straightforward, literalistic fashion, as if the Bible were a modern science or history textbook. The Bible wasn't written to answer our questions about *mechanisms*—*how* God built our universe. It's meant to tell us about *meaning*—*why* God built our universe, and what our purpose is within it.

Influenced by many great Christian thinkers, I've come to understand that the Bible is not primarily a science textbook, or a list of rules, or a collection of newspaper articles. It is primarily a *story*—a story of a lost, wayward, rebellious people and a God who, through Jesus, would stop at nothing to bring them home. This is the overarching story of the Bible, the story that underlies all the smaller stories. Christianity's bold claim is that this is also the True Story of the world.

Today, I no longer read about God's judgment through Noah's flood, in which God destroys the wicked but spares the righteous, without thinking about the story of Jesus, in which God destroyed the righteous to spare the wicked. I don't read the Old Testament laws condemning adulterers to death by stoning without also picturing Jesus, standing beside a woman caught in adultery in defiance of that very law, daring whoever thinks they are without sin to cast the first stone. And I don't read the ceremonial cleanliness laws without thinking of Jesus touching the lepers, making himself ceremonially unclean in order to cleanse them of their sickness and their sin.

I am grieved by the ways that the Bible has been used throughout history to justify oppression, hatred, and violence, but I am heartened that Jesus, in his short time on earth, stood against all such evils time and time again. C. S. Lewis put it like this: "It is Christ Himself, not the Bible, who is the true word of God. The Bible, read in the right spirit and with the guidance of good teachers will bring us to Him."[19]

Not long ago, I was visiting Dad in his Southern California apartment when Steve called. Steve was taking a course on the Old Testament as training for his job, and as a home-work assignment, he was supposed to call a family member or friend, share some thoughts on a passage from the Old Testament, and ask for their feedback. Steve picked a passage in which God promises Abraham that he would bless him and his descendants so that they, in turn, could bless the world. Steve talked about how Abraham is included in the genealogy of Jesus, suggesting that God ultimately fulfilled his promise to bless the world generations later through Abraham's great, great, great, great . . . great-grandson Jesus.

"What do you think of that?" Steve asked, as he concluded his short message. Dad was skeptical that God had actually spoken audibly to Abraham, and in light of that, it seemed like he had trouble tracking with the rest of Steve's thoughts on the passage. I held my tongue until Steve exchanged "I love yous" with both of us and hung up.

"What did you think of that?" Dad asked me.

"Well, I'm not sure about all of Steve's interpretation," I said, "but I definitely agree that a lot of the Bible makes a lot more sense in light of the story of Jesus. For example, I think the story of God calling Abraham to sacrifice his son Isaac sounds strange and nonsensical until you realize that whereas God stopped short of forcing Abraham to sacrifice his son Isaac, God *did* ultimately sacrifice his own son, Jesus."

"Huh," said Dad.

In all the years we had debated theology, it was the first time I had ever elicited anything but disagreement from him.

Epilogue

WHEN STEVE BEGAN INTRODUCING ME to the God of the Bible in 2009, I realized that what I had been rejecting my whole life was a caricature of God, a counterfeit of the real thing. Looking back, I'm amused by how little I knew about the arguments for and against God when I first became a Christian. Thirteen years later, I realize that I have only just begun to understand God's grandeur. Nevertheless, before I close, I feel that I should give a bit more explanation of what I believe today, and why.

As a kid, I found science to be trustworthy and useful. I still do today, but more than anything, I find it *astounding*. It boggles my mind that advanced mathematics, such as complex analysis, Lie theory, and algebraic geometry, play a role in the fundamental laws of nature. No matter your religious background, it's amazing that math so accurately and elegantly describes the world around us. It's incredible that science works at all.

To me, the mathematical coherence of the laws of nature rules out the possibility that our universe simply exists as a brute fact. There must be a more fundamental reality that explains the "unreasonable effectiveness" of mathematics to describe the natural world.[1] In the words of Aron Wall, a fellow Christian and theoretical physicist, "Either (1) the fundamental reality is something a bit like a mathematical equation (yet not a mere abstraction, but something which actually makes the world go around), or (2) the fundamental reality is something a bit like a mathematician, i.e., a mind capable of appreciating mathematical relations."[2]

If I were an atheist, I would advocate the first option: Our universe is nothing more than a mathematical structure, with laws that not only *describe* reality, but actually *define* reality. Yet, if I were an atheist, I would be bothered by the fact that so many features of our universe—and of the human psyche—are unexplained if our world is nothing but math in motion.

Our universe features a flow of time, a notion of cause and effect, conscious minds (namely, ours), and meaningful interactions between conscious minds—we are not isolated, self-aware beings who popped into existence. We conscious beings seem to possess free will, moral duties, purpose in life (if nothing more than to do good rather than evil), emotions, the ability to think rationally and seek out truth in the world around us, and a sense of beauty in the world around us. These features are so familiar to us as humans that we tend to forget how incredible they are.

To me, these features suggest that our world is better viewed not as a mathematical structure but as a *story*, presumably told by a cosmic storyteller. (It stands to reason that a mind intelligent enough to comprehend the equations of string theory and breathe them into life would also be intelligent enough to tell a story, and to breathe life into the characters of that story.)

Yet the hypothesis that our world is a great, cosmic narrative is not without a few problems. For one thing, the narrative we inhabit seems to lack any sort of point, in and of itself. In the words of Bertrand Russell, "All the labors of the ages, all the devotion, all the inspiration, all the noonday brightness of human genius, are destined to extinction in the vast death of the solar system, and . . . the whole temple of man's achievement must inevitably be buried beneath the debris of a universe in ruins."[3]

Following Russell's line of thinking, the human story lacks a climax: a moment when all hope seems lost, yet somehow, against all odds, good triumphs and evil defeats itself.

The human story needs a hero, someone who overcomes great obstacles, is tempted and tried, yet in the climactic moment, remains uncorrupted, risking his or her life for the good of the many.

This is the narrative arc of every great story ever told.

And this is where Jesus enters the story and the historical evidence for Christianity comes into its own. Jesus is the hero that the human story demands, the one who conquers not merely a mortal antagonist but rather the greatest enemy of all: *Death*. In his crucifixion, Jesus remains uncorrupted by evil, yet Death defeats itself, and in the climactic moment of history Jesus rises victorious over the grave, ensuring the happily-ever-after that every great story deserves.

The Christian story is a story worthy of a world, a story great enough to merit elevation from logical possibility to ontological reality, and a story that answers the age-old questions, "Why am I here?" and "What is the purpose of my life?"

If our world is truly a great cosmic narrative, then the existence of evil and suffering is not so much a problem as a feature. *Every* great story has evil and suffering; indeed, the greatest and happiest stories are often the ones with the *most* evil and suffering, because the worse things are at their worst, the more joyous the victory when it finally comes.

Even before I was a Christian, I was perpetually living out a story in my head. But in that story, *I* was always the hero, the one around whom the story revolved. As a kid, I dreamed of becoming a famous baseball player, dominating the opposing team with my arsenal of pitches. When I got to college, I dreamed of working for the NSA, using my mind to change the world of global politics. Even when I first came to faith, I dreamed of becoming a famous apologist, crushing opposing debaters with the brilliance of my arguments. In every one of my wildest fantasies of the future, I stood alone on the mountaintop of success.

In the end, I found that all these stories led only to loneliness. My faith provided a better story, one with meaning, hope, and purpose. And perhaps more than anything else, it made me part of a community, even a family.

Today, my dreams consist of telling jokes around a dinner table with my small group Bible study. I look forward to the Sunday morning doorbell, announcing that my next-door neighbors are ready to walk to church and are inviting me to tag along. And I dream of the day when I will get a text from Steve inviting me to Chicago for my nephew's baptism.

On that day, just as Steve's Christian friends did decades before at Northwestern, I will drink a root beer to celebrate. I will tell corny jokes about imaginary books of the Bible. And I will give my nephew a big hug and tell him I love him.

As long as he promises not to become Ned Flanders.

Notes

FOREWORD

1. Genesis 50:20, ESV.

INTRODUCTION

1. For a good discussion of the intersection of science, reason, and religion, see Ian Hutchinson, *Monopolizing Knowledge: A Scientist Refutes Religion-Denying, Reason-Destroying Scientism* (Belmont, MA: Fias, 2011).

PART I: THE ROAD TO FAITH

1. George Carlin, *You Are All Diseased*, TV special, HBO, February 6, 1999, Cable Stuff Productions.
2. Romans 14:8, NIV.
3. For a discussion of the origin of this saying, see "Absence of Evidence Is Not Evidence of Absence," *Quote Investigator* (blog), September 17, 2019, https://quoteinvestigator.com/2019/09/17/absence/#r+436457+1+4.
4. NIV.
5. Matthew 9:10-13.
6. Matthew 15:29-31.
7. Matthew 8:1-3.
8. Matthew 23:13-15, NIV.
9. Matthew 15:1-9.
10. Matthew 23:37.
11. James Allan Francis, *One Solitary Life* (Naperville, IL: Simple Truths, 2005), 61. [https://archive.org/details/onesolitarylife0000kenb/page/60/mode/2up?q=obscure+village.]
12. Galatians 1:13-14, NIV.
13. Philippians 2:6, NIV.

248 ≈ CHASING PROOF, FINDING FAITH

14. 2 Corinthians 11:24-27, esv.
15. All quotes here are from Richard Dawkins, *The God Delusion* (Boston: Houghton Mifflin, 2006), 157–158.
16. Kai Nielsen, *Reason and Practice* (New York: Harper & Row, 1971), 144. Italics in the original.
17. Peter Lipton, *Inference to the Best Explanation*, second edition (London: Routledge, 1991), 22.
18. Dawkins, *The God Delusion*, 158.
19. This formulation is derived from Quentin Smith, "Simplicity and Why the Universe Exists," *Philosophy*, vol. 72 (1997), 125–132. Smith writes: "The Law of the Simplest Beginning says that the simplest possible thing, the big bang singularity, comes into existence in the simplest possible way. The simplest possible way for something to come into existence is for the thing's coming into existence to have no positive relations to and grounds for coming into existence. The simplest possible way to come into existence is to come to exist *from nothing* (from no previously existent material, no material cause), to come to exist *by nothing* (by no efficient cause) and to come to exist *for nothing* (for no purpose or final cause). If the Law of the Simplest Beginning is true, then the big bang singularity occurs without being caused by God." Italics in the original.

PART II: NO SUCH AGENCY

1. Philippians 1:21, niv.
2. C. S. Lewis, *Mere Christianity*, first HarperCollins paperback edition (New York: HarperCollins, 2001), 56.
3. Timothy Keller with Kathy Keller, *The Meaning of Marriage: Facing the Complexities of Commitment with the Wisdom of God* (New York: Dutton, 2011).
4. For more information about United to Beat Malaria, see beatmalaria.org.
5. Luke 19:8.
6. Luke 19:9.
7. niv.
8. Justice Potter Stewart, concurring opinion in Jacobellis v. Ohio, 378 U.S. 184 (1964).
9. *Rush Hour*, New Line Cinema, 1998. Story by Ross LaManna, screenplay by Jim Kouf and Ross LaManna, directed by Brett Ratner.

PART III: FAITH AND DOUBT

1. Helmut Koester, *Introduction to the New Testament, Volume Two: History and Literature of Early Christianity* (Philadelphia: Fortress Press, 1982), 84.

2. Paula Fredriksen, *Jesus of Nazareth, King of the Jews: A Jewish Life and the Emergence of Christianity* (New York: Vintage Books, 2000), 264.
3. Bart D. Ehrman, *The New Testament: A Historical Introduction to the Early Christian Writings*, fourth edition (New York: Oxford University Press, 2008), 244.
4. Ehrman, 240–244.
5. Gary Habermas, Antony G. N. Flew, *Did Jesus Rise From the Dead? The Resurrection Debate*, ed. Terry L. Miethe (San Francisco: Harper & Row, 1987), 7.
6. Carl Sagan, *Cosmos* (New York: Random House, 1980).
7. Augustine, *St. Augustine, the Literal Meaning of Genesis.* vol. 1, Ancient Christian Writers, vol. 41, trans. John Hammond Taylor (New York: Paulist Press, 1982), 41.
8. Job 38:1–40:2, author's summary.
9. Job 38:24-27.
10. C. S. Lewis, *A Grief Observed* (New York: HarperOne, 1961, 1994), 6–7.

PART IV: THE KING
1. Mark 10:43-45.
2. C. S. Lewis, *The Screwtape Letters*, first HarperCollins paperback edition (New York: HarperOne, 2001), 40.
3. John 6:55.
4. NIV.
5. "Where Is Evidence of God, Tom Rudelius," TJump podcast, https://music.amazon.fr/podcasts/2af75e0e-dc47-4d72-b692-6778de4aa82c/episodes/348b2f43-43b7-45c6-bb12-d6e6bcb6ee28/tjump-where-is-evidence-of-god-tom-rudelius.
6. Timothy Keller, lecture at Massachusetts Institute of Technology, Cambridge, MA, February 26, 2008.
7. John R. W. Stott, *The Cross of Christ*, 20th anniversary edition (Downers Grove, IL: IVP, 1986, 2006), 326–327. John Stott quotes the final sentence, "The cross of Christ . . . is God's only self-justification in such a world," from P. T. Forsyth, *The Justification of God* (London: Duckworth, 1916), 32.
8. Fyodor Dostoevsky, *The Brothers Karamazov*, trans. Constance Garnett (London: William Heinemann, 1912), 241.
9. Matthew 12:39.
10. Marcus J. Borg, "The Mighty Deeds of Jesus," beliefnet, https://www.beliefnet.com/faiths/christianity/2004/04/the-mighty-deeds-of-jesus.aspx.
11. Matthew 12:24.
12. 1 Corinthians 13:12.

13. Matthew 5:38-39.

14. Matthew 5:43-44.

15. Matthew 19:8-9.

16. Jacob Neusner, *A Rabbi Talks with Jesus: An Intermillennial, Interfaith Exchange* (New York: Doubleday, 1993), 30–31.

17. Rachel Held Evans, *Inspired: Slaying Giants, Walking on Water, and Loving the Bible Again* (Nashville, TN: Nelson, 2018), 11–12.

18. Evans, xx.

19. C. S. Lewis, letter to Mrs. Johnson, November 8, 1952, in *The Collected Letters of C. S. Lewis, Volume III: Narnia, Cambridge and Joy 1950–1963*, ed. Walter Hooper (New York: HarperCollins, 2007), 246.

EPILOGUE

1. "The Unreasonable Effectiveness of Mathematics in the Natural Sciences" is a famous article written by physicist Eugene Wigner, published in *Communications in Pure and Applied Mathematics*, vol 13, no. 1 (February 1960).

2. Aron Wall, "Fundamental Reality V: Some Candidates, and a Math Test," *Undivided Looking* blog, December 20, 2014, http://www.wall.org/~aron /blog/fundamental-reality-v-some-candidates-and-a-math-test/. Wall continues, "(I don't mean either of these descriptions to be taken too literally here, obviously the fundamental entity cannot be exactly like a set of symbols on the blackboard, or a human mind, but the choice of analogy makes the difference to what effects seem likely to follow.) If the former hypothesis is true, we would have *Naturalism*, a worldview which takes the universe as revealed by the Natural Sciences to be the ultimate reality, so that everything else must depend on that. If the latter is true, we would have a *Supernatural* or *Theistic* view of reality. I don't mean to suggest that (1) and (2) are the only possible candidates for the fundamental entity, just the ones I find most plausible."

3. Bertrand Russell, "A Free Man's Worship," in *The Meaning of Life: A Reader*, ed. E. D. Klemke and Steven M. Cahn (Oxford: Oxford University Press, 2008), 56.

About the Author

TOM RUDELIUS completed his undergraduate work at Cornell, earned a doctorate in physics at Harvard, and has conducted postdoctoral research at the Institute for Advanced Study in Princeton, New Jersey. Currently a postdoctoral researcher in theoretical physics at the University of California, Berkeley, Tom will begin a faculty position at Durham University in the fall of 2023. His research focuses on string theory, quantum field theory, and early universe cosmology. A man of faith and an avid sports fan, he is frequently asked to speak on topics related to science and faith. He is also on the board of the Mamelodi Initiative, a tutoring organization based in Mamelodi, South Africa.

TYNDALE
REFRESH™

Think Well. Live Well. Be Well.

Experience the flourishing of your mind, body, and soul with Tyndale Refresh.